Aldol Reactions

Rainer Mahrwald

Aldol Reactions

Springer

Rainer Mahrwald
Institute of Chemistry
Humboldt University
Brook Taylor Str. 2
12 489 Berlin
Germany
rainer.mahrwald@rz.hu-berlin.de

ISBN: 978-1-4020-8700-4 e-ISBN: 978-1-4020-8701-1
DOI: 10.1007/978-1-4020-8701-1
Springer Dordrecht Heidelberg London New York

Library of Congress Control Number: 2009922617

Cover illustration: Photographs by the author

Cover design: WMXDesign GmbH Heidelberg

Printed on acid-free paper

Springer is part of Springer Science+Business Media (www.springer.com)

Preface

Without doubt, the aldol reaction belongs to the most important methods of stereoselective C-C bond formation processes. One of the requirements of modern synthetic methods is receiving chiral products in their enantiomerically pure form or pure diastereoisomers. This requirement is not only important for synthetic chemistry but also an imperative for nature. The aldol reaction fits in easily with nature's chemistry. A great number of enzymatic transformations are based on the aldol addition. This fact has been known for a long time and was best expressed with the following statement:

Nature, it seems, is an organic chemist having some predilection for aldol condensation.
J. W. Cornforth in Perspectives in Organic Chemistry, Todd, A. Ed.; Wiley-Interscience, New York, 1956, page 371.

The stormy and meteoric development in the field of aldol additions over the last 20 years has led to a plethora of options in the total syntheses of complicated natural products. This book illustrates the basic principals and the new variants of the classical aldol addition. This includes aldol additions with various metal enolates as well as metal complex-catalyzed, organocatalytic methods and biocatalytic transformations. Moreover, advances made in the development of new aldol methodologies to provide pure stereoisomers are preferentially discussed in this work.

The exciting progress being made in organocatalytic aldol additions is of particular interest. This highly active topic of research will continue to develop an increasing number of new concepts of configuration-control. This idea is perfectly illustrated by the following scheme – something unimaginable 10 years ago. This scheme brings together organocatalyzed aldol additions and enzymatic transformations, which flow perfectly into the idea of Cornforth mentioned above.

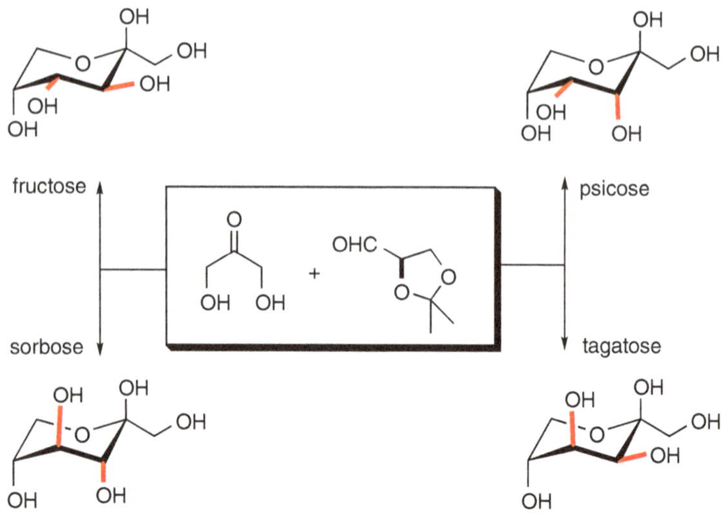

Using four different organocatalysts, an approach to four differently configured ketohexoses is possible today. In nature, on the other hand, a selective and asymmetric access to the described ketohexoses is accomplished by a family of four aldolases. These enzymes selectively target one of the four isomers of the 1,2-diol junction that link dihydroxyacetone with (*R*)-glyceraldehyde. Thus, an approach is given selectively to one of the four enantiomers by biochemical as well as organocatalytic methods.

It is now my pleasure to express my gratitude to my co-workers and my son for countless hours of assistance. Special thanks are also due to Springer UK, especially to Miss Claudia Thieroff and Dr. Sonia Ojo.

Humboldt-University, Berlin
March 2009

Rainer Mahrwald

Contents

List of Abbreviations

9-BBN	9-Borabicyclo[3.3.1]nonane
ab	Antibody
Ac	Acetyl
AcOH	Acetic acid
ALB	Aluminium–lithium–bis[(S)-binaphthoxide] complex
Ar	Aryl
BDMS	Benzyldimethylsilyl
BINAP	2,2′–bis(diphenylphosphino)–1,1′–binaphthyl
BINAPO	2,2′–bis(diphenylphosphinoxide)–1,1′–binaphthyl
BINOL	1,1′-bi-2-naphthol
Bn	Benzyl
Boc	*tert*-butoxycarbonyl
BOM	Benzyloxymethyl
box	Oxazoline
BPS	*tert*-butyldiphylsilyl
Bu	Butyl
Bz	Benzoyl
c	Cyclo
CAB	Chiral acyloxyboranes
Cbz	Benzyloxycarbonyl
COD	Cyclooctadien
Cp	Cyclopentadienyl
CSA	Camphorsulfonic acid
Cy	Cyclohexyl
DBU	1,8-Diazabicyclo[5.4.0]undec-7-ene
DHA	Dihydroxyacetone
DHAP	Dihydroxyacetone phosphate
DIPEA	Diisopropylethylamine
DMSO	Dimethyl sulfoxide
Dppe	2-(Diphenylphosphino)ethyl
Et	Ethyl

FBP	Fructose-l,6-bisphosphate
FruA	Fructose-l,6-bisphosphate-aldolase
FucA	L-Fuculose-1-phosphate-aldolase
GABOB	α-Amino-β-hydroxybutanoic acid
HMDS	1,1,1,3,3,3-Hexamethyldisilazane
HMG-CoA	3-Hydroxy-3-methylglutaryl-coenzyme A
HMPA	Hexamethylphosphoramide
HYTRA	Hydroxy-1,1,2-triphenyl ethyl acetate
Ipc	Isopinocampheyl
LDA	Lithium-diisopropylamine
LLB	LLB Lanthanium–lithium–BINOL
Me	Methyl
MEM	2-Methoxyethoxymethyl
Mes	Methanesulfonyl
MOM	Methoxymethyl
MS	Molsieve
NAL	N-Acetylneuraminic acid lyase
NeuA	N-Acetylneuraminic acid aldolase
Ph	Phenyl
PMB	p-Methoxybenzyl
PMP	p-Methoxyphenyl
Pr	Propyl
p-TsOH	p-Toluenesulfonic acid
pybox	Pyridyl bis(oxazoline)
RhuA	L-Rhamnulose-1-phosphate-aldolase
RibA	2-Deoxy-D-ribose 5-phosphate aldolase
TagA	Tagatose-1,6-bisphosphate aldolase
TBAF	Tetrabutylammonium fluoride
TBAI	Tetrabutylammonium iodide
TBDPS	t-Butyldiphenylsilyl
TBS	t-Butyldimethylsilyl
TCE	1,1,1-Trichloroethyl
TES	Triethylsilyl
Tf	Trifluoromethanesulfonate
TFA	Trifluoroacetic acid
THF	Tetrahydrofuran
Thr	Threonine
TIPS	Triisopropylsilyl
TMEDA	N,N,N′,N′-Tetramethylethylenediamine

TMS	Trimethylsilyl
Tol-BINAP	2,2′–bis(di-p-toluylphosphino)–1,1′–binaphthyl
TPS	Triphenylsilyl
Tr	Triphenylmethyl (trityl)
TRAP	(R,R)-2,2′-bis[(S)-1-(dialkylphosphanyl)ethyl]1,1′-biferrocene
Troc	2,2,2-Trichloroethoxycarbonyl
Ts	p-Toluenesulfonyl

1 Introduction

Aldol reactions have been reviewed several times as a specific topic in an exhaustive manner under different aspects.[1,2,3,4,5,6,7,8,9]

Due to the stormy development in this field of organic chemistry, especially in the last 10 years, there is a need for giving an update and summarizing the existing differing aldol methodologies. This is of particular importance in the field of organocatalysed aldol addition.

One of the requirements of modern synthetic methods is the production of chiral substances in their enantiomerically pure form or pure diastereoisomers. Therefore, advances made in the development of new aldol methods, which provide pure stereoisomers, will be discussed here preferentially. There are, in general, three different main strategies to achieve this goal:

- Addition of chiral enolates to aldehydes (including both the use of chiral ketones and chiral auxiliaries)
- Addition of achiral enolates to chiral aldehydes
- Use of chiral metal ligands, chiral catalysts (metals as well as organocatalysts), chiral additives or enzymes

Several aspects are responsible for the configurative outcome of metal enolate-driven aldol additions. One of the most important features is the architecture of the enolate used in these reactions. The correlation between enolate geometry and configuration of the aldols obtained is important for the evaluation of the existing transition state models, which have been applied so successfully for trisubstituted enolates. A chair-like transition state proposed by Zimmerman and Traxler[9] has been well accepted to account for the correlation of the (*E*)- or (*Z*)-geometry of the enolate to the *syn*- or *anti*-configuration in the aldol product (Scheme 1.1). Although first developed for magnesium enolate addition in the Ivanoff reaction[10] this model was successfully applied to stereochemical outcomes of aldol additions even of boron and titanium enolates. Transition state **A** (R^1 equatorial) is expected to have a lower energy than transition state **B** (R^1 axial) in the (*Z*)-enolate series. As a consequence of this consideration a preference of *syn*-configured aldol adducts **1** is obtained. In the (*E*)-enolate series transition state **C** (R^1 equatorial) is favoured over the transition state **D** (R^1 axial). Formation of *anti*-configured aldol adducts should then be preferred.

R. Mahrwald, *Aldol Reactions*,

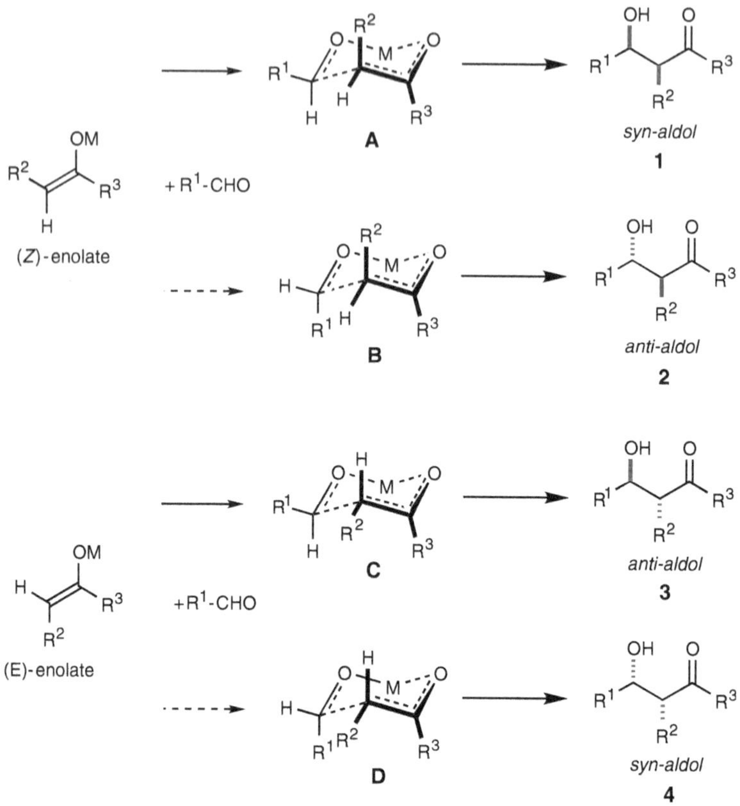

Scheme 1.1 Zimmerman–Traxler transition state models in aldol additions of (*E*)- and (*Z*)-enolates

When used with different counter ions different configurative results in aldol additions can be observed (Scheme 1.2). This can be explained by several additional transition state models like skewed transition states, boat transition states, twist-boat transition states, 1,3-dipolar cycloaddition transition states and open transition states. For a discussion see also Braun.[11]

Modern aldol methods based on the use of preformed enolates were developed as clean reactions with high yields in order to prevent side reactions. It turned out that during this development several metals proved to be key players in applications of preformed enolates. These are lithium, magnesium, boron, tin and titanium. The generation and their application will be discussed in this book. Scheme 1.2 represents a very impressive and instructive example concerning the stereochemical outcome of the application of these four metals in aldol additions. Based on preformed enolates of lithium, boron, magnesium or titanium the four stereoisomeric aldol adducts **6–9** of chiral ketone **5** can be obtained in a controlled manner.

The different stereochemical outcomes of these aldol additions can be explained by a chelate transition state **E** (lithium enolates) and by a non-chelate transition state **F** (boron enolates) (Scheme 1.3).[12]

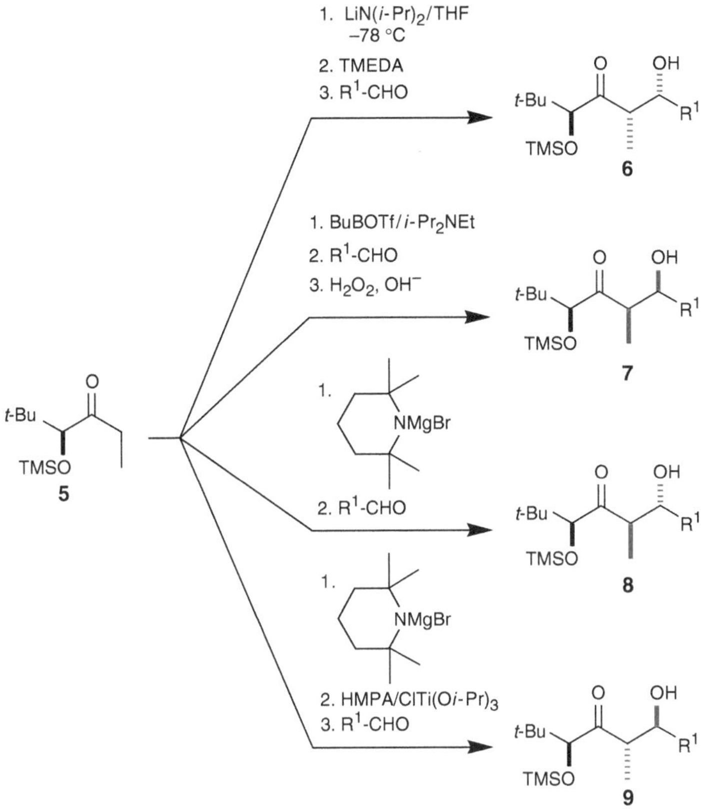

Scheme 1.2 Stereodivergence in propionate aldol additions

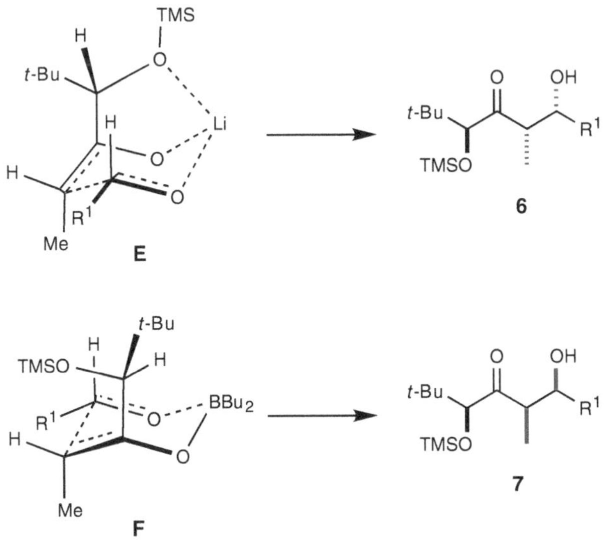

Scheme 1.3 Chelate and non-chelate transition state models

Another impressive example is the diastereoselective aldol addition of chiral β-ketoamides as shown in Scheme 1.4. When used with titanium, boron or tin enolates an access to different configured aldol adducts **11**, **12** and **13** is given.[13,14] Explanations for these configurative outcomes are given by transition state models **G**, **H** and **I**. A further illustrative example of diastereoselection in propionate aldol addition is shown in Scheme 1.5. Again, *anti*-configuration was observed (for aldol adduct **18**) when boron enolates of chiral ketone **15** were reacted with chiral aldehyde **14**.[15]

These few examples should be sufficient in the introduction to illustrate the power and diversity of different aldol methodologies to install required configurations during the C–C bond formation. Several applications of different aldol methods demonstrate this in total synthesis of natural products. The pros and cons of application of metal enolates or catalysed aldol additions (Lewis acid, Lewis base, organo- and enzyme-catalysis) will be discussed in this book. Furthermore, comparisons will be given with regard to stereoselective outcomes using different aldol methods.

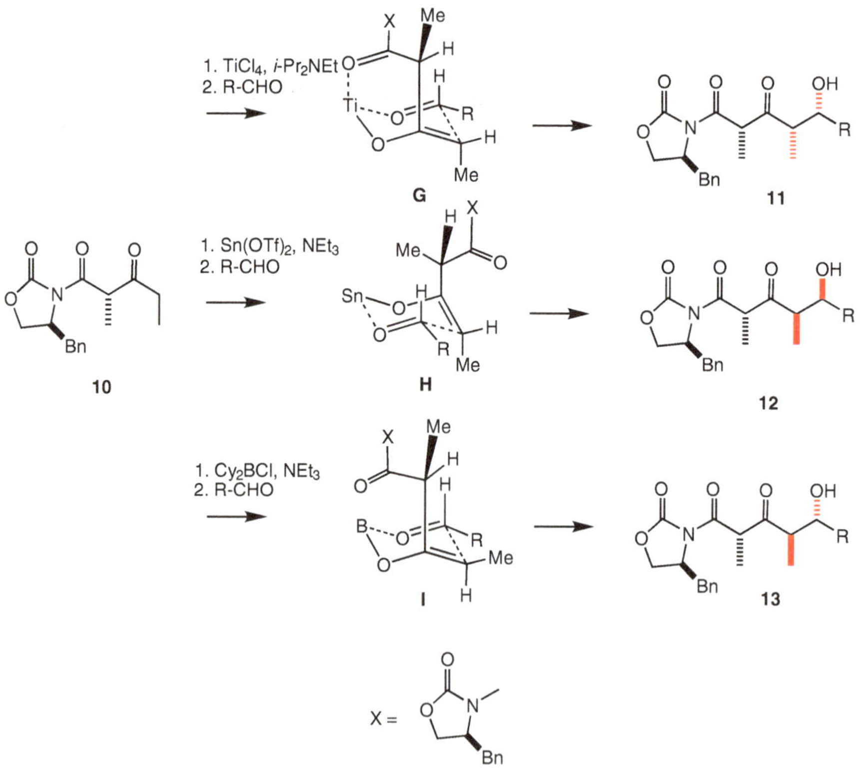

Scheme 1.4 Stereoselection in propionate aldol additions

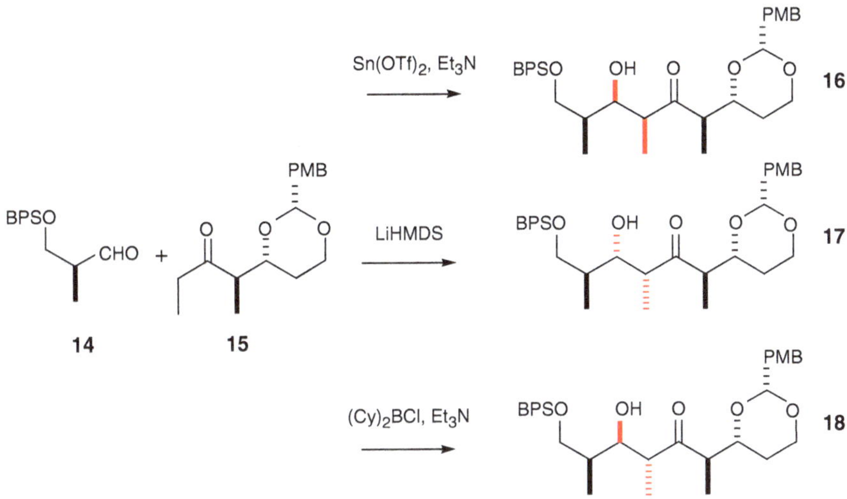

Scheme 1.5 1.2-Asymmetric induction in synthesis of extended propionate fragments

References

1. Paterson I, Franklin AS (1994) Contemp Org Synth 1:317
2. Braun M (1996) In: Houben-Weyl, Methoden der Organischen Chemie, Helmchen G, Hoffmann RW, Mulzer J, Schaumann E (eds). Thieme, Stuttgart. vol E21b, p 1603
3. Machajewski TD, Wong CH (2000) Angew Chem In Ed 39:1352
4. Palomo C, Oiarbide M, Garcia JM (2002) Chem Eur J 8:36
5. Palomo C, Oiarbide M, Garcia JM (2004) Chem Soc Rev 33:65
6. Mahrwald R (ed) (2004) Modern Aldol Reactions. Wiley-VCH, Weinheim
7. Schetter B, Mahrwald R (2006) Angew Chem Int Ed 45:7506
8. Schetter B, Mahrwald R (2008) In: Science of Synthesis, Clayden J (ed). Thieme, Stuttgart, p 847
9. Geary LM, Hultin PG (2009) Tetrahedron: Asymm 20:131
10. Ivanoff D, Spassoff A (1931) Bull Soc Chim Fr 49:371
11. Braun M (2004) In: Modern Aldol Reactions, Mahrwald R (ed). Wiley-VCH, Weinheim, vol 1, p 1
12. van Draanen NA, Arseniyadis S, Crimmins MT, Heathcock CH (1991) J Org Chem 56:2499
13. Evans DA, Clark JS, Metternich R, Novack VJ, Sheppard GS (1990) J Am Chem Soc 112:866
14. Evans DA, Burch JD, Hu E, Jaeschke G (2008) Tetrahedron 64:4671
15. Menche D, Arikan, F, Li J (2008) Org Lett 10:3521

2 Aldol Reactions with Preformed Enolates

2.1 Lithium Enolates

The synthesis of lithium enolates and their application in aldol additions have been the subject of several reviews.[1,2,3,4,5,6] These are highly reactive, their handling is easy and they can be used on a large scale – even on an industrial scale.[7] Ideal starting compounds proved to be chiral ene components. For that reasons most published results used this route. Heathcock et al. used lithium enolates of chiral carbohydrate-derived ketones in aldol additions. By additions to aldehydes moderate diastereoselectivities were detected.[8,9,10] Pioneering investigations were made by Seebach and coworkers.[11] Enantiomerically pure 3-methyl-2-pentanone was converted into the corresponding lithium enolate. Subsequently addition of acetaldehyde, propionaldehyde or benzaldehyde yielded the expected β-hydroxycarbonyl compounds. Later on, Seebach and coworkers developed the concept of 'self-reproduction of chirality', which is based on the use of chiral lactones.[12] Aldehydes and unsymmetrical ketones were added to lithium enolates of readily available chiral acetales derived from lactic acid, mandelic acid or amino acids.[13,14,15] High stereoselectivities were achieved. Liebeskind and Davies demonstrated that optically active iron acyl complexes can serve as chiral ene components.[16,17] Thus, through a diastereoselective reaction high stereoselectivities were observed. An improvement of this strategy was achieved by introducing a pentafluorophenyl containing phosphane ligand instead of triphenylphosphane.[18] Due to acceptor–donor interactions of enolate oxygen and perfluorinated phenyl ring high stereoselectivities in reactions with aldehydes were observed. Yamamoto and coworkers applied acetates to aldol additions containing an axial chirality. The lithium enolates react with aldehydes in a highly stereoselective manner.[19] Braun et al. developed a concept based on the use of hydroxy-1,1,2-triphenylethyl acetate (HYTRA).[20] The starting materials – both enantiomers of methyl mandelate – are inexpensive and readily available. Double deprotonation of the starting chiral acetate **1** (commercially available) and addition to aldehydes yielded aldol adducts **2** and **3** in high diastereomeric ratios.[21,22,23,24] The diastereoselectivity can be enhanced by further adding magnesium halides (see Scheme 2.1.1).

The reliability of this transformation was demonstrated by the application in the synthesis of a large number of biologically active compounds as well as natural products. This corresponds to the syntheses of γ-amino-β-hydroxybutanoic acid (GABOB),[25] shikonin and alkannin,[26] digitoxose,[27] detoxinine[28] and statin.[29] Even stereoselective syntheses of tetrahydrolipstatine,[30] compactin,[31] epothilones,[32] (23*S*)-hydroxyvitamin

R. Mahrwald, *Aldol Reactions*,

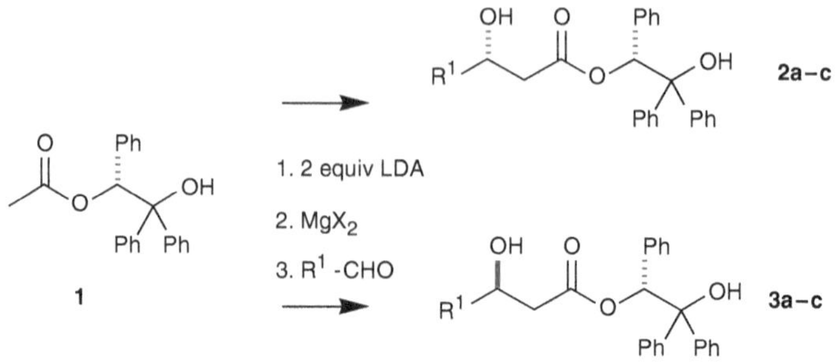

R^1	MgX_2 (mol%)	ratio **2 / 3**
R^1: Ph	0 mol % MgX_2	**2a / 3a** - 83 / 17
R^1: Ph	200 mol % $MgBr_2$	**2a / 3a** - 88 / 12
R^1: Ph	100 mol % MgI_2	**2a / 3a** - 98 / 2
R^1: Et	100 mol % MgI_2	**2b / 3b** - 96 / 4
R^1: *n*–Pr	200 mol % $MgBr_2$	**2c / 3c** - 93 / 7

Scheme 2.1.1 Influence of magnesium halides in HYTRA aldol additions

D3 derivatives[33] and synthetic inhibitors of HMG-CoA reductase[34] were carried out on an industrial scale with the aid of HYTRA aldol methodology.

The addition of lithium enolates to chiral aldehydes proved to be the second access to optically active β-hydroxycarbonyl compounds. A very early and prominent example of this strategy is illustrated by the introduction of the C1–C2 segment of erythromycin A. The group of Woodward published this synthesis in 1981(Scheme 2.1.8).[35] Heathcock and coworkers reported the results of aldol additions using lithium enolates and chiral α-alkoxy-substituted aldehydes, such as isopropylidene glyceraldehyde.[36]

Aldol additions of lithium enolates to N-protected α-amino aldehydes yielded the *anti*-configured adducts. A systematic study was carried out by Reetz and coworkers and is shown in Scheme 2.1.2.[37]

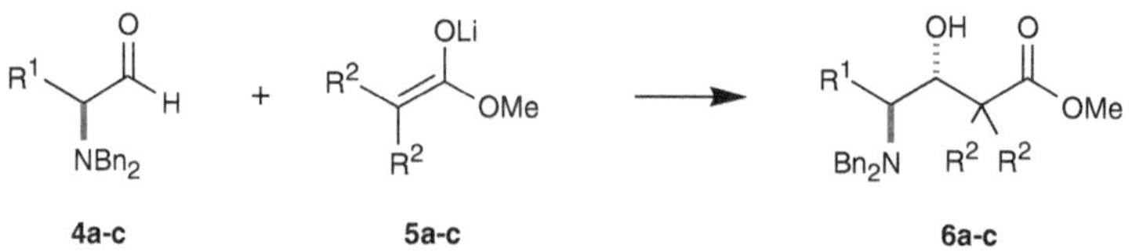

	yield (%)	*anti / syn*
6a, R^1 = Me, R^2 = H	82	95 / 5
6b, R^1 = Bn, R^2 = H	83	90 / 10
6c, R^1 = Me, R^2 = Me	84	97 / 3

Scheme 2.1.2 *anti*-Selective aldol additions of N-protected α-amino aldehydes

For stereoselection in the N-protected α-amino ketone series see Lagu and Liotta.[38] A comparative and systematic study of enolate additions with α-alkoxy-substituted aldehydes and α,β-alkoxy-disubstituted aldehydes was carried out by Evans and coworkers (Scheme 2.1.3). They observed that asymmetric induction in additions of lithium enolates to α-alkoxy-substituted aldehydes is superior to the results obtained by additions of boron enolates or enol silyl ethers (Scheme 2.1.4).[39]

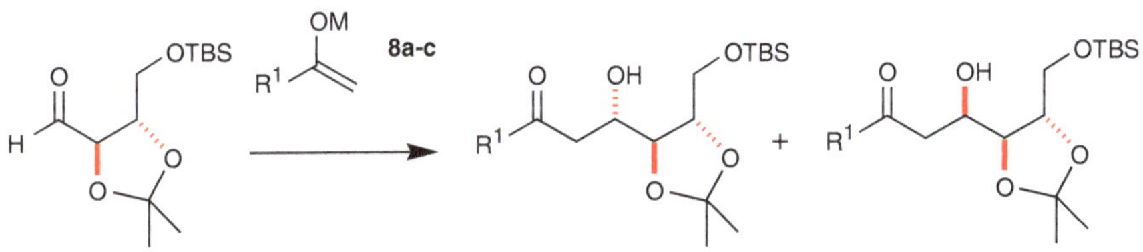

entry	R^1	ratio of **9a-c / 10a-c** (yield in %)		
		M = TMS/$BF_3 \cdot OEt_2$	M = 9-BBN	M = Li
1	**a**, Me	65 / 35 (61)	49 / 51 (74)	89 / 11 (62)
2	**b**, *i*-Pr	67 / 33 (69)	54 / 46 (91)	92 / 8 (48)
3	**c**, *t*-Bu	54 / 46 (28)	54 / 46 (72)	90 / 10 (58)

Scheme 2.1.3 Aldol reactions of *syn*-α,β-bisalkoxy aldehydes

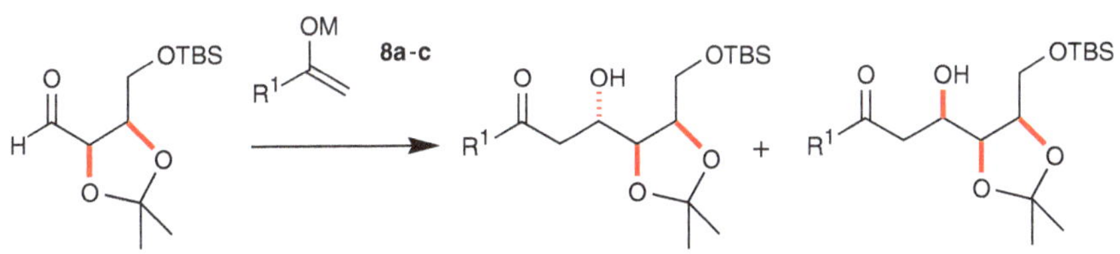

entry	R^1	ratio of **12a-c / 13a-c** (yield in %)		
		M = TMS/$BF_3 \cdot OEt_2$	M = 9-BBN	M = Li
1	**a**, Me	83 / 17 (81)	67 / 33 (98)	86 / 14 (81)
2	**b**, *i*-Pr	63 / 37 (61)	76 / 24 (85)	96 / 4 (85)
3	**c**, *t*-Bu	25 / 75 (69)	78 / 22 (74)	94 / 6 (90)

Scheme 2.1.4 Aldol reactions of *anti*-α,β-bisalkoxy aldehydes

Only a few examples were published where chiral lithium reagents were used successfully in stereoselective aldol additions (Schemes 2.1.5 and 2.1.6).[40] This access is the field of boron, titanium and other metal enolates. Nevertheless, some aldol reactions of chiral lithium amide are shown in Scheme 2.1.5 and 2.1.6.

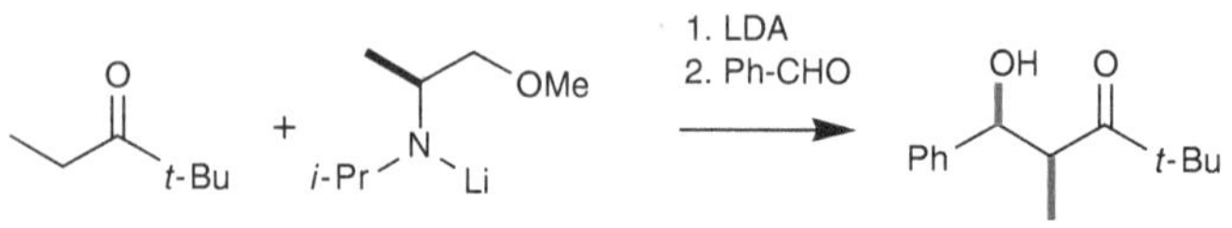

14 **15** **16:** 92%; 68% ee

Scheme 2.1.5 Use of chiral lithium amide in an enantioselective aldol addition

The generation of chiral enolates of protected dihydroxyacetone was carried out in the presence of chiral lithium amide. High *anti*-selectivities were detected during these aldol reactions. In some cases aldol adducts were obtained with a high degree of enantioselectivity (Scheme 2.1.6).[41]

Nakamura and coworkers demonstrated that the configurative outcome of aldol additions of trisubstituted enolates does not depend on the identity of the metal atom (**28**: $R^Z \neq H$, Scheme 2.1.7).[42] This result stands in contrast to those obtained in aldol additions of disubstituted metal enolates, among which *E*-enolates ($R^Z = H$) sometimes react via a boat transition state **B** leading to products **30**. Destabilization of the latter in the presence of R^Z of persubstituted enolates is probably the reason for this experimental observation.

The stereoselective use of trisubstituted lithium enolates is limited to those reactions in which a rapid *syn–anti* equilibration of the aldol products via lithium enolates does not take place (entries 1 and 3, Scheme 2.1.7). Otherwise, only poor selectivities were obtained (entry 2, Scheme 2.1.7).

Very instructive examples of the use of lithium enolates in polyketide synthesis can be found in the total synthesis of erythronolide A and epothilones. Woodward and coworkers used lithium enolates of *tert*-butyl thiopropionate for the introduction of the C1–C2 unit of the erythronolide A seco acid **36**, as mentioned above [35]. In this way, they exclusively obtained product **35**, which still contains the undesired configuration at C2. Compound **36** containing the 'natural' configuration at C2 was obtained by kinetic protonation of compound **35** (Scheme 2.1.8).

Schinzer and coworkers used the corresponding chiral acetonide **39** as lithium enolate source. This was the key step in the total synthesis of epothilones (Fig. 2.1.1, Scheme 2.1.9).[43,44] The aldol addition of acetonide **39** with the chiral aldehyde (*S*)-**41** resulted in the formation of **42** as the major isomer (25:1). An explanation of this remarkable stereochemical result is given by the transition state shown in Scheme 2.1.9. Two functional peculiarities of the acetonide **39** lead to this chelation-controlled model: firstly, the influence of the tertiary α-carbon atom generating (*Z*)-enolate **40**, and secondly, the oxygen functionality at C3 leading to that rigid bicyclic transition state structure shown in Scheme 2.1.9. For further development of total syntheses of epothilones see also Nicolaou and Montagnon.[45]

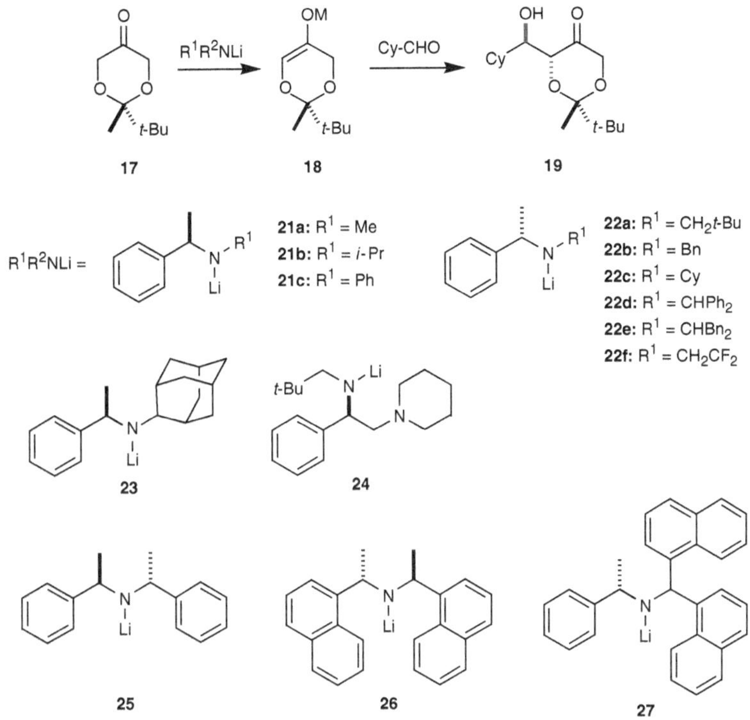

entry	lithium amide	LiCl (equiv)	ee (%)	yield (%)
1	**21a**	0	(–)10	43
2	**21b**	0.5	(–)16	55
3	**21c**	0.5	(–)32	53
4	**22a**	1	(+)19	63
5	**22b**	0.5	(+)39	56
6	**22c**	0.5	(+)20	51
7	**22d**	1	(+)70	60
8	**22e**	1	(+)60	76
9	**22f**	1	(+)90	61
10	**23**	1	(+)80	91
11	**24**	0.5	(+)20	51
12	**25**	0.5	(+)60	51
13	**26**	0.5	(–)60	43
14	**27**	1	(+)90	95

Scheme 2.1.6 Yields and enantioselectivities of *anti*-configured aldol adduct **19**

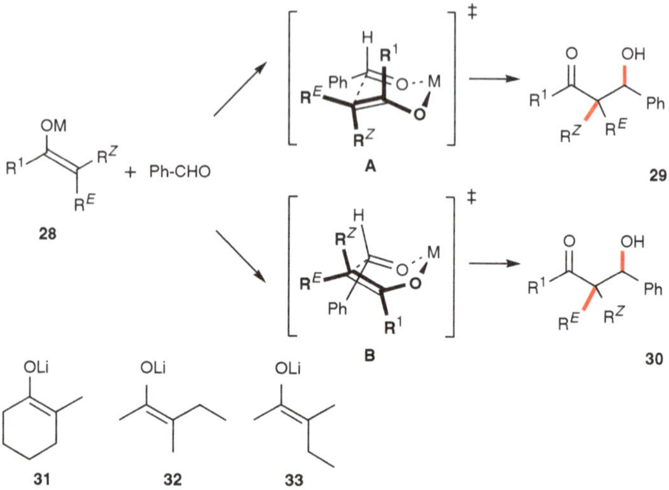

entry	enolate	**29 / 30**	yield (%)
1	**31**	9 / 91	86
2	**32** (97%*Z*)	45 / 55	76
3	**33** (86%*E*)	17 / 83	71

Scheme 2.1.7 Aldol additions of trisubstituted lithium enolates to benzaldehyde

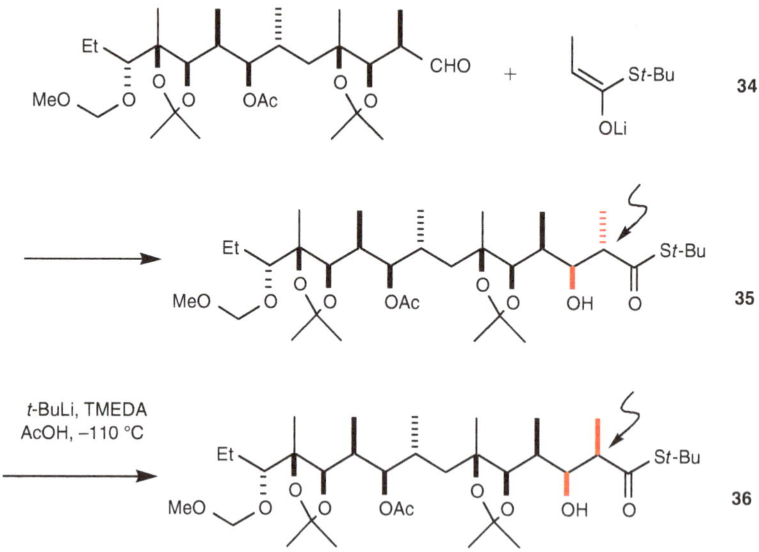

Scheme 2.1.8 Total synthesis of the aglycon of erythronolide A

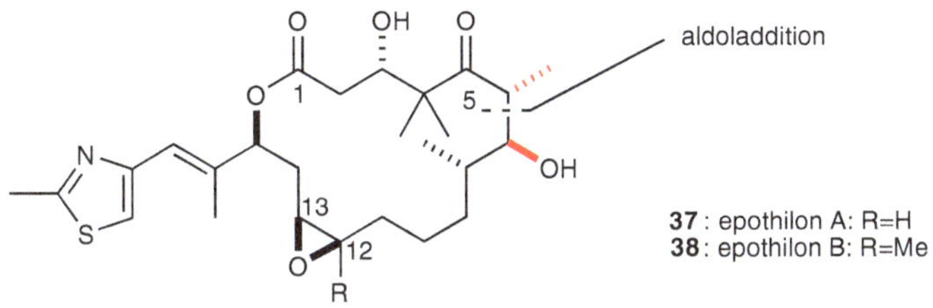

Fig. 2.1.1 Natural occurring epothilones

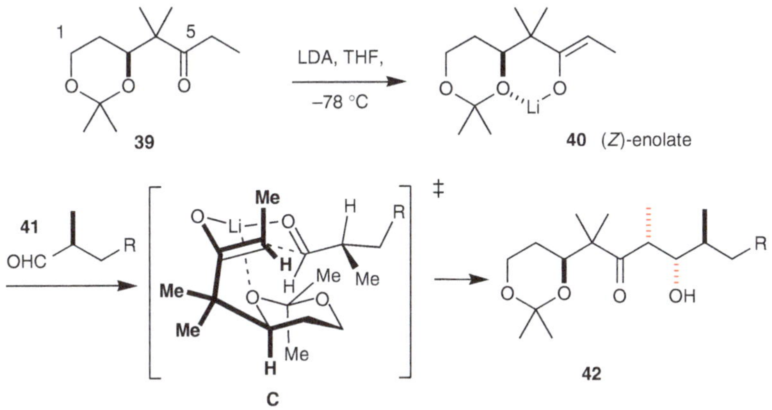

Scheme 2.1.9 Total synthesis of epothilon by Schinzer and coworkers

Further applications of lithium enolates in aldol additions were reported in the total synthesis of spongistatin 2.[46] A comparison of configurative outcome of boron with lithium enolates in the aldol coupling of AB- and CD-spiroacetale subunits **43** and **44** is given in Scheme 2.1.10.

For the deployment of lithium enolates in total synthesis of pheromones see also Pilli et al.[47]

Lithium enolates were often used for further transmetallation in order to increase stereoselectivities. During the total synthesis of spiruchostatin A the optically active intermediate **47** was needed. The lithium enolate of Seebach's *N*-acetyl-oxazolidin-2-one **45**[48] was used in an acetate aldol addition. Enhancement of diastereoselectivity could be achieved by transmetallation of the lithium enolate (Scheme 2.1.11).[49]

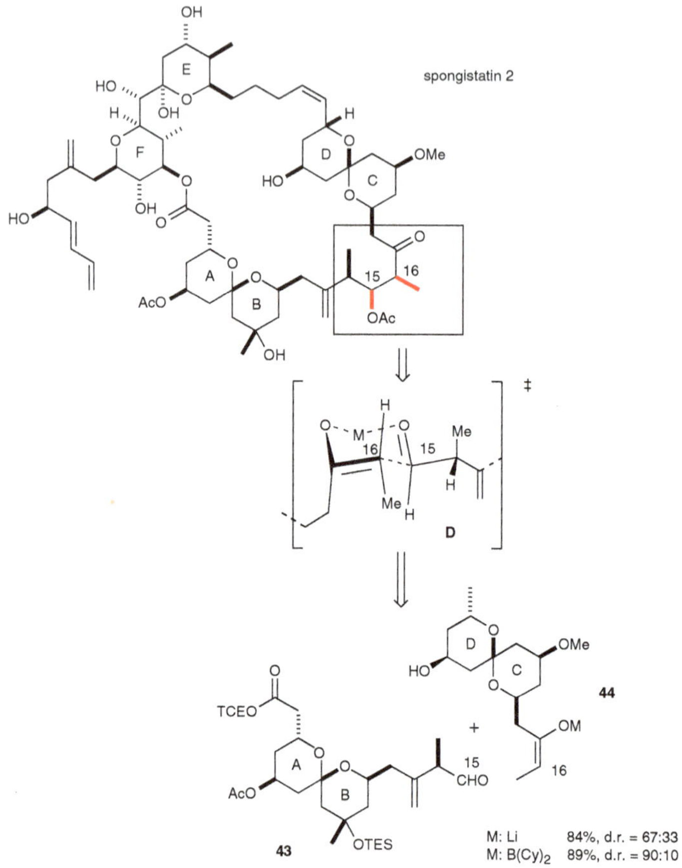

Scheme 2.1.10 Total synthesis of spongistatin 2

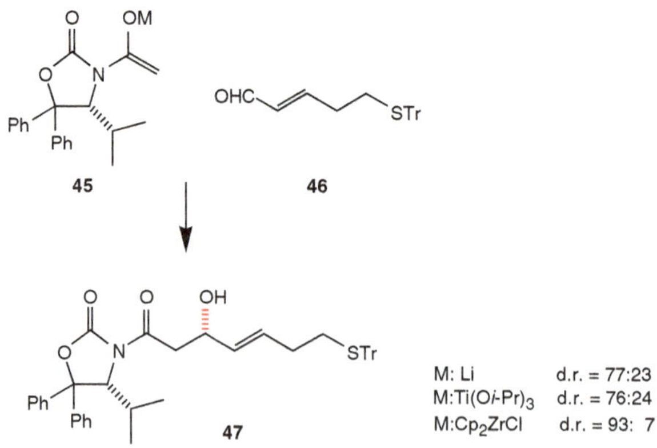

Scheme 2.1.11 Total synthesis of spiruchostatin A

References

1. Meckelburger HB, Wilcox CS (1993) In: Comprehensive Organic Synthesis, Trost BM, Fleming I, Heathcock CH (eds). Oxford, Pergamon, vol 2, p 99
2. Heathcock CH (1993) In: Comprehensive Organic Synthesis, Trost BM, Fleming I, Heathcock CH (eds). Oxford, Pergamon, vol 2, p 181
3. Heathcock CH (1992) In: Modern Synthetic Methods, Scheffold R (ed). Verlag Sauerlaender, Aarau, p 1
4. Braun M (2004) In: Modern Aldol Reactions, Mahrwald R (ed). Wiley-VCH, Weinheim, vol 1, p 1
5. Braun M (2004) In: The Chemistry of Organolithium Compounds, Rappoport Z, Marek I (eds). Wiley, Chichester, p 829
6. Tomooka K, Ito M (2004) In: Main Group Metals in Organic Synthesis, Yamamoto H (ed). Wiley-VCH, Weinheim, vol 1, p 1
7. (a) Totter F, Rittmeyer P (1994) In: Organometallics in Synthesis – A Manual, Schlosser M (ed). Wiley, Chichester, p 167; (b) Hodgson D (2003) Organolithiums in Enantioselective Synthesis. Springer, New York
8. Heathcock CH, White CT, Morrison JJ, vanDerveer D (1981) J Org Chem 46: 1296
9. Heathcock CH, Pirrung MC, Buse CT, Hagen JP, Yound SD, Sohn JE (1979) J Am Chem Soc 101:7077
10. Lodge EP, Heathcock CH (1987) J Am Chem Soc 109:3353
11. Seebach D, Ehrig V, Teschner M (1976) Liebigs Ann Chem 1357
12. For a comprehensive overviews see (a) Seebach D, Sting AR, Hoffmann M (1996) Angew Chem 108:2880; (b) Seebach D, Imwinkelried R, Weber T (1986) In: Modern Synthetic Methods Scheffold R (ed). Verlag Sauerlaender, Aarau, p 125
13. Seebach D, Naef R, Calderari G (1984) Tetrahedron 40:1313
14. Seebach D, Weber T (1984) Helv Chim Acta 67:1650
15. Amberg W, Seebach D (1990) Chem Ber 123:2413
16. Davies SG, Dordor IM, Warner P (1984) Chem Commun 956
17. For a comprehensive overview see McCallum K, Liebeskind LS (1996) In: Houben-Weyl, Methoden der Organischen Chemie, Helmchen G, Hoffmann RW, Mulzer J, Schaumann E (eds). Thieme, Stuttgart, vol E21b, p 1667
18. Ojima I, Kwon HB (1988) J Am Chem Soc 110:5617
19. Keito SS, Hatanka K, Kano T, Yamamoto H (1998) Angew Chem Int Ed 37:3378
20. Braun M (1987) Angew Chem Int Ed Engl 26:24
21. Braun M, Dervant R (1984) Tetrahedron Lett. 25:5031
22. Dervant R, Mahler U, Braun M (1988) Chem Ber 121:397
23. Braun M, Gräf S, Herzog S (1993) Org Synth 72:32
24. Braun M, Gräf S (1993) Org Synth 72:38
25. Braun M, Waldmüller S (1989) Synthesis 856
26. Braun M, Bauer C (1991) Liebigs Ann Chem 1157
27. Braun M, Moritz J (1991) Synlett 750
28. Ewing WR, Harris BD, Bhat KL, Joullie MM (1986) Tetrahedron 42:2421
29. Wuts PGM, Putt SR (1989) Synthesis 951
30. (a) Barbier P, Schneider U, Widmer U (1987) Helv Chim Acta 70:1412; (b) (1986) Eur Patent 185359
31. (a) Lynch JE, Volante RP, Wattley JV, Shinkai I (1987) Tetrahedron Lett. 28:1385; (b) (1988) Eur Patent 251714
32. for an overview of this development see Schinzer D (2004) In: Modern Aldol Reactions, Mahrwald R (ed). Wiley-VCH, Weinheim, vol 1, p 311
33. Yamanashi SR, Shimizu N (1966) Jpn Patent 08301811
34. (a) Patel DV, Schmidt RJ, Gordon EM (1992) J Org Chem 57:7143; (b) German Patent 3805801

35. Woodward RB, Logush E, Nambiar KP, Sakan K, Ward DE, Au-Yeung BW, Balaram P, Browne LJ, Card PJ, Chen CH, Chênevert RB, Fliri A, Frobel K, Gais HJ, Garratt DJ, Hayakawa K, Heggie KW, Hesson DP, Hoppe D, Hoppe I, Hyatt JA, Ikeda D, Jacobi PA, Kim KS, Kobuke Y, Kojima K, Krowicki K, Lee VJ, Leutert T, Malchenko S, Martens J, Matthews RS, Ong BS, Press JB, Rajan Babu TV, Rousseau G, Sauter HM, Suzuki M, Tatsuta K, Tolbert LM, Truesdale EA, Uchida I, Ueda Y, Uyehara T, Vasella AT, Vladuchick WC, Wade PA, Williams RM, Wong HNC (1981) J Am Chem Soc 103:3210
36. Heathcock CH, Young SD, Hagen JP, Pirrung CT, White CT, vanDerveer J (1980) J Org Chem 45:3846
37. Reetz MT, Drewes MW, Schmitz A (1987) Angew Chem Int Ed Engl 26:1141; for a comprehensive overview see also Reetz MT (1999) Chem Rev 99:1121
38. Lagu BR, Liotta DC (1994) Tetrahedron Lett 35:4485
39. Evans DA, Cee VJ, Siska SJ (2006) J Am Chem Soc 128:9433
40. Ando A, Shiori T (1989) Tetrahedron 45:4969
41. (a) Majewski M, Nowak P (2000) J Org Chem 65:5152; (b) Majewski M, Ulaczyk-Lesanko A, Wang F (2006) Can J Chem 84:257
42. Yamago S, Machii D, Nakamura E (1991) J Org Chem 56:2098
43. Schinzer D, Bauer A, Bohm OM, Limberg A, Cordes M (1999) Chem Eur J 5:2483
44. Schinzer D, Bauer A, Schieber J (1999) Chem Eur J 5:2492
45. Nicolaou KC, Montagnon T (2008) Molecules That Changed the World. WILEY-VCH, Weinheim
46. Paterson I, Chen DYK, Coster MJ, Acena JL, Bach J, Wallace D (2005) Org Biomol Chem 3:2431
47. Pilli RA, Murta MM, Russowsk D, Boeckelmann MA (1991) J Braz Chem Soc 2:121
48. Hintermann T, Seebach D (1998) Helv Chim Acta 81:2093
49. Doi T, Iijima Y, Shin-a K, Ganesan A, Takahashi T (2006) Tetrahedron Lett. 47:1177

2.2 Magnesium Enolates

There are only a few examples of applications of magnesium enolates in aldol additions. A very early example of a successful employment of magnesium enolates in stereoselective aldol additions was reported by Solladie and coworkers.[1,2,3] Deprotonation of sulfinyl acetate with *tert*-butylmagnesium bromide yielded aldol adducts with moderate to high enantioselectivities (Scheme 2.2.1).

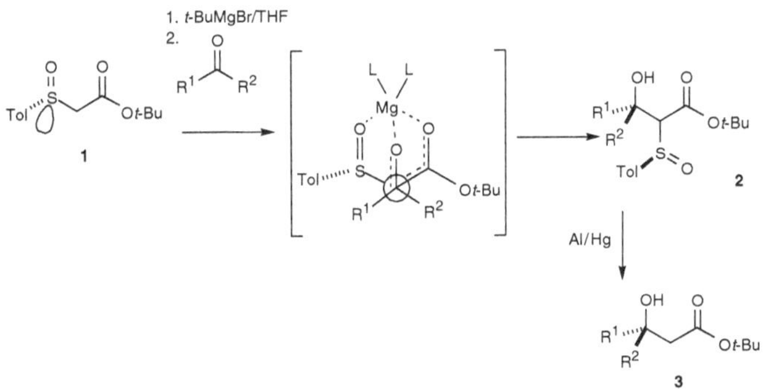

entry	compound	yield (%)	ee (%)
1	**3a**, R^1 = H, R^2 = C_7H_{15}	80	86
2	**3b**, R^1 = H, R^2 = $C{\equiv}CC_3H_7$	73	80
3	**3c**, R^1 = H, R^2 = $C{\equiv}CC_6H_{13}$	53	70
4	**3d**, R^1 = H, R^2 = Ph	85	91
5	**3e**, R^1 = Me, R^2 = C_7H_{15}	88	95
6	**3f**, R^1 = Me, R^2 = $C{\equiv}CC_3H_7$	60	48
7	**3g**, R^1 = Me, R^2 = $C{\equiv}CC_6H_{13}$	72	36
8	**3h**, R^1 = Me, R^2 = CO_2Et	80	8
9	**3i**, R^1 = Me, R^2 = $C_2H_4O_2CMe$	90	40
10	**3k**, R^1 = Me, R^2 = Ph	75	68
11	**3l**, R^1 = Ph, R^2 = CF_3	75	20

Scheme 2.2.1 Enantioselective aldol addition of sulfinyl acetates

R. Mahrwald, *Aldol Reactions*,

Very recently, an extension of this method to the synthesis of chiral *syn*- and *anti*-configured 1,2-diols was published.[4]

But the application of Grignard reagents in direct aldol additions is limited to special substrates as well as to special Grignard reagents. It causes side reactions, such as the reduction of aldehydes.

More applications were found by the use of magnesium amides – as an alternative to lithium amides in synthesis.[5,6] Henderson and coworkers described aldol additions performed in the presence of magnesium-hexamethyldisilazane.[7] They were able to elaborate a synthesis of crystalline bis-(hexamethyldisilazido)-magnesium. Moreover, they were also able to isolate a magnesium intermediate from the self aldol addition of methyl *tert*-butyl ketone. The structure of this intermediate was determined by x-ray structure analysis. An application of a sterically hindered magnesium amide is reported by Heathcock and coworkers.[8] But even the application of magnesium amides caused disadvantages like reduction of aldehydes as one can see by low yields in the enolizable aldehyde series.

The most important application of magnesium compounds in aldol additions represents the use of magnesium salts. They can be used in

- Direct aldol additions in combination with amine bases or as
- Transmetallating agents added to preformed lithium enolates or as
- Lewis acids in Mukaiyama reactions.

Braun and Dervant described the use of $MgBr_2$ and MgI_2 in aldol additions of lithium enolates of HYTRA. A transmetallation is suspected to be responsible for the significant improvement of the stereoselectivity observed.[9]

A systematical study of employment of magnesium salts in aldol additions was carried out by Liotta and coworkers (Scheme 2.2.2).[10]

These investigations indicate a thermodynamical equilibration yielding the *anti*-configured aldol adducts with a high degree of stereoselectivity. A similar effect of thermodynamical control by using magnesium compounds in aldol additions were described by Hayashi et al.[11] and Mahrwald.[12]

The application of magnesium salts in combination with amine bases was reported. Nagao et al. described aldol additions of bislactim ethers with enolizable aldehydes in the presence of magnesium bromide and triethylamine.[13] A more general application was reported by Pare and coworkers. They described in a series of publications the use of magnesium iodide in the presence of equimolar amounts of amine bases.[14,15,16] The aldol adducts of non-enolizable aldehydes were observed with high degrees of *anti*-diastereoselectivity (Scheme 2.2.3).

For the use of magnesium salts in aldol additions of silyl enol ethers see Fujisawa et al.[17] An application of magnesium perchlorate in the synthesis of α,β-dihydroxyketones was published by Willis et al.[18] Aldol adducts of aldehydes and thioesters can easily be obtained in the presence of magnesium bromide and tertiary amines.[19,20]

Stereochemical investigations in $MgBr_2$-catalysed direct aldol additions have been carried out by Evans and coworkers. They described *anti*-selective direct aldol additions of chiral *N*-acyloxazolidinones **10** with non-enolizable aldehydes

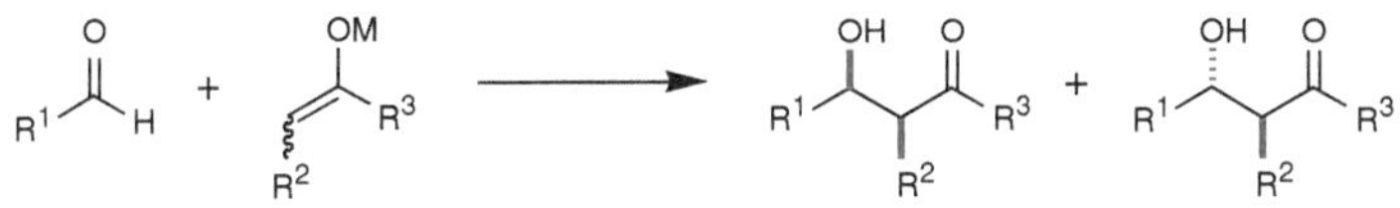

4 5 6 7

entry	R^1	R^2	R^3	*anti / syn* (yield, %)[a]		time (h)[a]
				LDA	$MgBr_2$	
1	Ph	Me	Ph	21 / 79	98 / 2 (73)	16
2	Ph	Me	Ph	21 / 79	73 / 27 (46)	16
3	*n*-C_5H_{11}	Me	Ph	n.r.	87 / 13 (65)	16
4	*n*-C_4H_9	Me	Ph	29 / 71	89 / 11 (54)	5
5	Ph	Me	Et	36 / 64	77 / 23 (68)	16
6	*i*-Pr	Me	Et	69 / 31	91 / 9 (74)	5
7	*i*-Pr	Et	Ph	69 / 31	95 / 5 (75)	5
8	Ph	Me	*i*-Pr	18 / 82	90 / 10 (78)	2.5
9	Ph	Me	*i*-Pr	18 / 82	91 / 9 (82)	22
10	*n*-C_5H_{11}	Me	*i*-Pr	21 / 72	81 / 19 (65)	16
11	Ph	Me	*t*-Bu	<2 / 98	99 / 1 (70)	6
12	*n*-C_5H_{11}	Me	*t*-Bu	<2 / 98	81/ 19 (50)	48
13	Ph	OMe	Ph	40 / 60	92 / 8 (86)	6

Scheme 2.2.2 *anti*-Selective aldol addition mediated by magnesium bromide
[a]For reactions with added $MgBr_2$

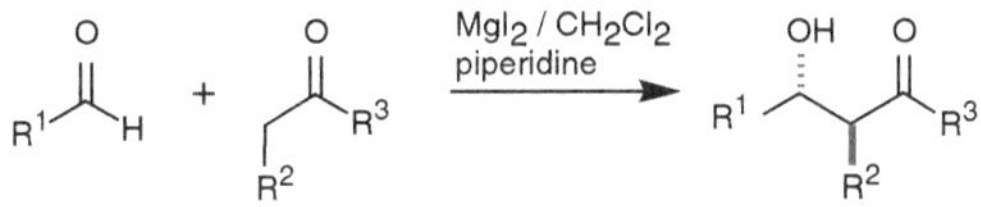

7

entry	R^1	R^2	R^3	*anti / syn*	yield (%)
1	Ph	Et	Ph	95 / 5	91
2	4-FC_6H_4	Et	Ph	94 / 6	92
3	4-BrC_6H_4	Et	Ph	90 / 10	90
4	4-MeC_6H_4	Et	Ph	92 / 8	88
5	$C_6H_5CH = CH$	Et	Ph	88 / 12	92
6	$CH_3CH = CH$	Et	Ph	84 / 16	90
7	*t*-Bu	Et	Ph	100 / 0	82
8	Ph	– $(CH_2)_4$–		77/ 23	68
9	4-ClC_6H_4	– $(CH_2)_4$–		76 / 24	70
10	Ph	Et	Et	76 / 24	70

Scheme 2.2.3 *anti*-Selective aldol addition mediated by magnesium iodide

catalysed by 10 mol% of $MgBr_2{\cdot}OEt_2$ (Scheme 2.2.5).[21] A short time afterwards they reported direct aldol additions under the same conditions using chiral *N*-acylthiazolidinethiones **8** and non-enolizable aldehydes (Scheme 2.2.4).[22]

The authors compared these two methods with regard to the different stereochemical outcomes. The unexpected *anti*-propionate adducts **9** obtained in the *N*-acylthiazolidinethiones series are complementary to adducts derived from reactions of chiral *N*-acyloxazolidinones **10**. These different stereochemical results can be explained by transition states **A** and **B** in Schemes 2.2.4 and 2.2.5.

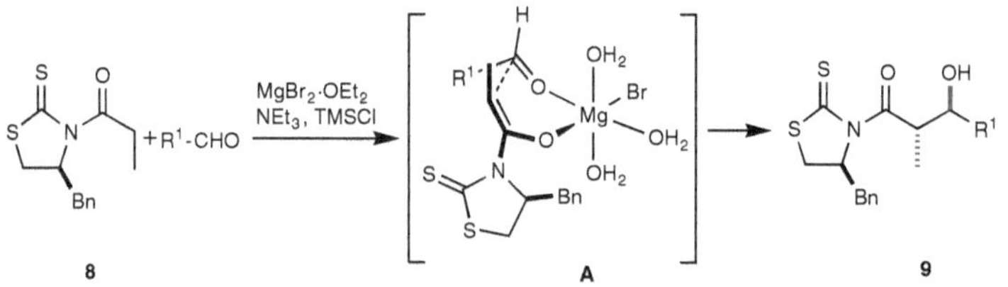

entry	R^1	*anti* / *syn*	yield (%)
1	Ph	95 / 5	85
2	4-MeC_6H_4	95 / 5	92
3	4-MeOC_6H_4	95 / 5	91
4	PhCH = CH	91 / 9	87
5	$PhCH = C(CH_3)$	95 / 5	90
6	2-naphthyl	88 / 12	84
7	$CH_2 = C(CH_3)$	88 / 11	56

Scheme 2.2.4 *anti*-Selective magnesium-mediated aldol additions of chiral *N*-acylthiazolidinethiones

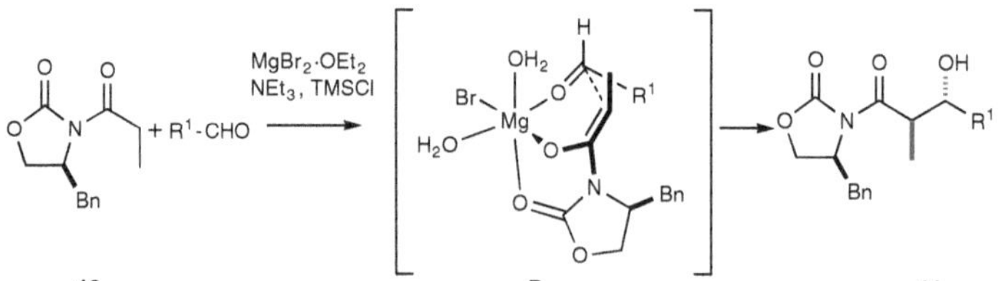

entry	R^1	*anti* / *syn*	yield (%)
1	Ph	97 / 3	91
2	4-MeC_6H_4	96 / 4	n.r.
3	4-MeOC_6H_4	97 / 3	91
4	PhCH = CH	95 / 5	92
5	PhCH = CMe	97 / 3	92
6	2-naphthyl	93 / 7	91

Scheme 2.2.5 *anti*-Selective magnesium-mediated aldol additions of chiral *N*-acyloxazolidinones

Recently McNulty and coworkers have extended this method to the use of enolizable aldehydes. High *anti*-selectivities were detected in aldol additions of chiral auxiliary **13** with protected L-threose **12**. Matched/mismatched situations were discussed by the use of both enantiomers of oxazolidinones. These results were utilized in total synthesis of pancratistatines (Scheme 2.2.6).[23,24]

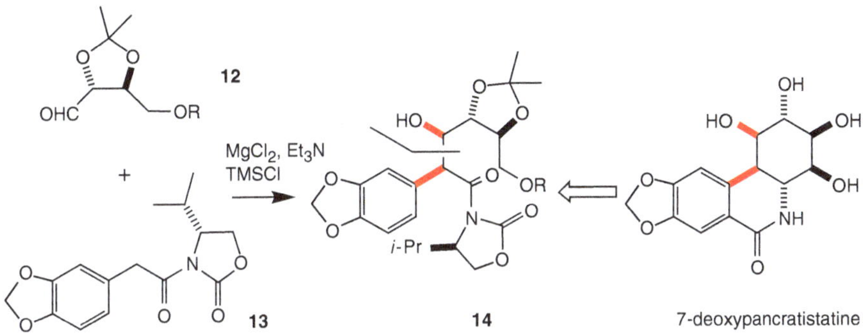

Scheme 2.2.6 Matched *anti*-aldol addition promoted by $MgCl_2$

The application of magnesium salts as Lewis acids in Mukaiyama reactions was studied by several groups.[25,26,27,28,29] The deployment of magnesium iodide in Mukaiyama reactions was reported by Corey and coworkers.[30] The authors described difficulties during a Mukaiyama aldol step in the total synthesis of lactacystin. These problems could be overcome by the deployment of MgI_2 in this Mukaiyama reaction step. The desired configuration of intermediate **15** was obtained by configurative specific chelation of magnesium iodide, which was confirmed by x-ray structure analysis (*anti*-**15**/*syn*-**15**: 9:1, Scheme 2.2.7).

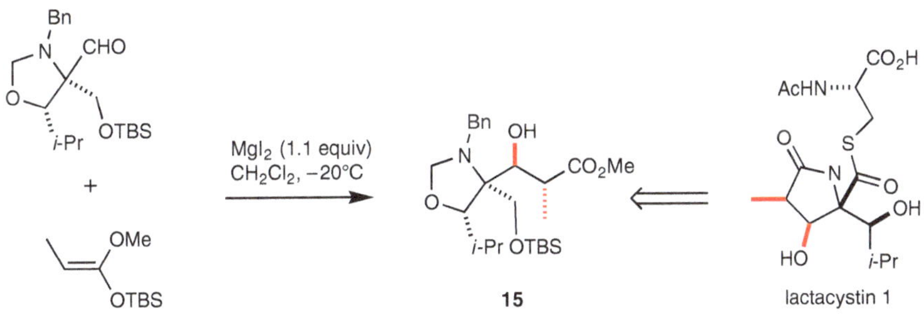

Scheme 2.2.7 Application of magnesium iodide in total synthesis of lactacystin

Successful application of magnesium bromide-mediated Mukaiyama reactions in a total synthesis of taxol was reported by Mukaiyama and coworkers.[31,32]

References

1. Mioskowski S, Solladie GJ (1977) Chem Soc 162
2. Mioskowski S, Solladie G (1980) Tetrahedron 36:227
3. Solladie G (1996) In: Houben-Weyl, Methoden der Organischen Chemie, Helmchen G, Hoffmann RW, Mulzer J, Schaumann E (eds). Thieme, Stuttgart, vol E21b, p 1808
4. Bauder C (2008) Tetrahedron Lett 49:2243
5. He X, Morris JJ, Noll BC, Brown SN, Henderson KW (2006) J Am Chem Soc 128:13599
6. He X, Allan JF, Noll BC; Kennedy AR, Henderson KW (2005) J Am Chem Soc 127:6920
7. Allan JF, Henderson KW, Kennedy AR (1999) Chem Commun 1325
8. vanDraanen NA, Arseniyadis S, Crimmins MT, Heathcock CH (1991) J Org Chem 56:2499
9. Braun M, Dervant R (1984) Tetrahedron Lett 25:5031
10. Swiss KA, Choi WB, Liotta D, Abdel Magid AF, Maryanoff CA (1991) J Org Chem 56:5978
11. Hayashi K, Kujime E, Katayama H, Sano S, Nagao Y (2007) Chem Pharm Bull 55:1773
12. Mahrwald R (1995) Tetrahedron 51:9015
13. Sano S, Miwa T, Liu XK, Shii T, Takehisa T, Shiro M, Nagao Y (1998) Tetrahedron: Asymm 9:3615
14. Wei HX, Li K, Zhang Q, Jasoni RL, Hu J, Pare PW (2004) Helv Chim Acta 87:2354
15. Wei HX, Hu J, Jasoni RL, Li G, Pare PW (2004) Helv Chim Acta 87:2359
16. Wei HX, Jasoni RL, Dchao H, Hu J, Pare PW (2004) Tetrahedron 60:11829
17. (a) Fujisawa H, Sasaki Y, Mukaiyama T (2001) Chemistry Lett 3:190; (b) Li WDZ, Zhang XX (2002) Org Lett 4:3485
18. Willis MC, Cutting GA, John MP (2004) Synlett 1195
19. Yost JM, Zhou G, Coltart DM (2006) Org Lett 8:1503
20. Zhou G, Yost JM, Coltart DM (2008) Synthesis 478
21. Evans DA, Tedrow JS, Shaw JT, Downey CW (2002) J Am Chem Soc 124:392
22. Evans DA, Downey CW, Shaw JT, Tedrow JS (2002) Org Lett 4:1127
23. McNulty J, Nair JJ, Sliwinski M, Harrington LE, Pandey S (2007) Eur J Org Chem 5669
24. McNulty J, Nair JJ, Griffin C, Pandey S (2008) J Nat Prod 71:357
25. Fujisawa H, Sasaki Y, Mukaiyama T (2001) Chem Lett 190
26. Takai K, Heathcock CH (1985) J Org Chem 50:3247
27. Uenishi J, Tomozane H, Yamato M (1985) Tetrahedron Lett 26:3467
28. Bernardi A, Cardani S, Colonbo L, Poli G, Schimperna G, Scolastico C (1987) J Org Chem 52:888
29. Li WDZ, Zhang XX (2002) Org Lett 4:3485
30. Corey EJ, Li W, Reichard GA (1998) J Am Chem Soc 120:2330
31. Mukaiyama T, Shiina I, Iwadare H, Saitoh M, Nishimura T, Ohkawa N, Sakoh H, Nishimura K, Tani Y, Hasegawa M, Yamada K, Saitoh K (1999) Chem Eur J 5:121
32. Shiina I, Shibata J, Ibuka R, Imai Y, Mukaiyama T (2001) Bull Chem Soc Jpn 74:113

2.3 Boron Enolates

Several reviews have been published to summarize the development of this very important method of aldol additions.[1,2,3,4,5,6] Nearly 30 years ago Mukaiyama et al. developed the fundamentals for this transformation.[7,8,9] After these initial reports the attention was drawn to the stereoselective execution of this method. The aldol addition proceeds via a chair-like, six-membered transition state, which is more rigid than those of alkali metal enolates. This is due to the shorter boron oxygen bond length, which guarantees a maximum of 1,3-diaxial interactions ($R^3 \leftrightarrow L$) and thus the formation of the more stable transition states **A** and **B**. For that reason higher stereoselectivities were observed when used with boron enolates, compared to aldol additions of corresponding lithium enolates. The stereochemical outcome strongly depends on the geometry of the boron enolates used in these reactions. (*E*)-Enolates provide the *anti*-configured aldol adducts, whereas *syn*-aldol adducts were formed by (*Z*)-enolates.[10] These results are illustrated by the transitions states shown in Scheme 2.3.1.

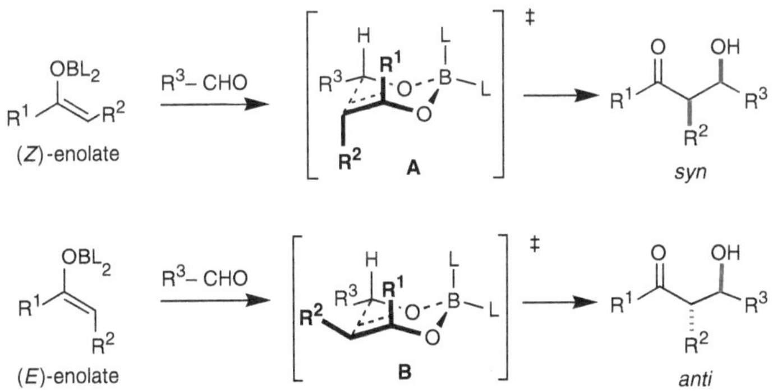

Scheme 2.3.1 Stereoselective aldol reactions of (*Z*)- or (*E*)-boron enolates and aldehydes

These considerations were confirmed by some examples illustrated in Scheme 2.3.2. For the synthesis of defined (*Z*)- or (*E*)-enolates see references.[11,12,13,14,15,16,17,18,19]

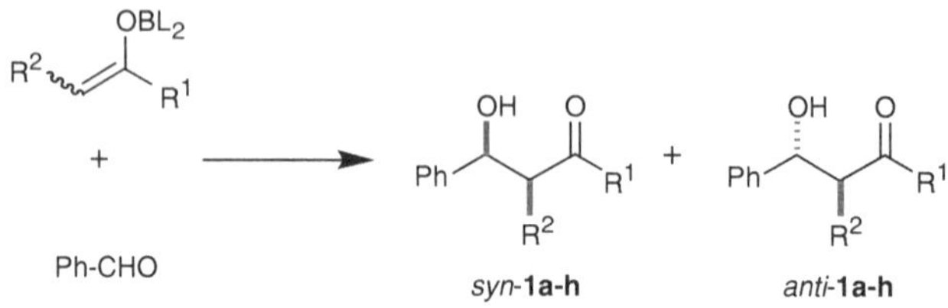

entry	compound	R^1	R^2	L	(*Z*) / (*E*)	*syn* / *anti*
1	**1a**	Me	Bu	Bu	100 / 0	>95 / 5
2	**1a**	Me	Bu	Bu	0 / 100	25 / 75
3	**1b**	Et	Me	Bu	>97 / 3	>97 / 3
4	**1c**	*i*-Pr	Me	*c*-C_5H_9	19 / 81	18 / 82
5	**1d**	Cy	Me	*c*-C_5H_9	12 / 88	14 / 86
6	**1e**	*t*-Bu	Me	Bu	>99 / 1	>97 / 3
7	**1f**	Ph	Me	Bu	99 / 1	>97 / 3
8	**1g**	S*t*-Bu	Me	Bu	5 / >95	10 / 90
9	**1h**	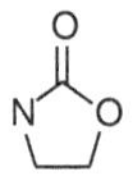	Me	Bu	>97 / 3	98 / 2

Scheme 2.3.2 Stereoselective aldol addition of different boron enolates to benzaldehyde

For *syn*-selective aldol additions using boron trichloride or alkoxydichloroborane in the presence of tertiary amines see Chow and Seebach.[20]

Currently (*E*)-boron enolates of ketones are mostly generated by using hindered dialkylboryl triflates and amines. The corresponding (*Z*)-enolates are prepared under kinetic-controlled conditions. Ketones are reacted with less-hindered boryltriflates at −78°C (Scheme 2.3.3).[21,22,23]

For a detailed protocol of generation and handling of boron enolates in aldol additions see Cergol and Coster.[24]

The same is true for the generation of defined boron enolates of carboxylic esters. (*E*)-boron enolates of carboxylic esters were obtained when used with *tert*-butyl esters and dicyclohexylboron triflates in the presence of triethylamine selectively. (*Z*)-Boron enolates of carboxylic esters were generated by the application of dibutylboron triflates and diisopropylethylamine to methyl or ethyl carboxylic esters (Scheme 2.3.3).[25]

As pointed out in the introduction there exist in general three different ways for an asymmetric execution of aldol reactions. Conventional chiral enolates have been used for this purpose extensively. Several chiral auxiliaries have been developed to achieve this aim. A selection of existing auxiliaries in boron enolate aldol additions is given in Scheme 2.3.4.

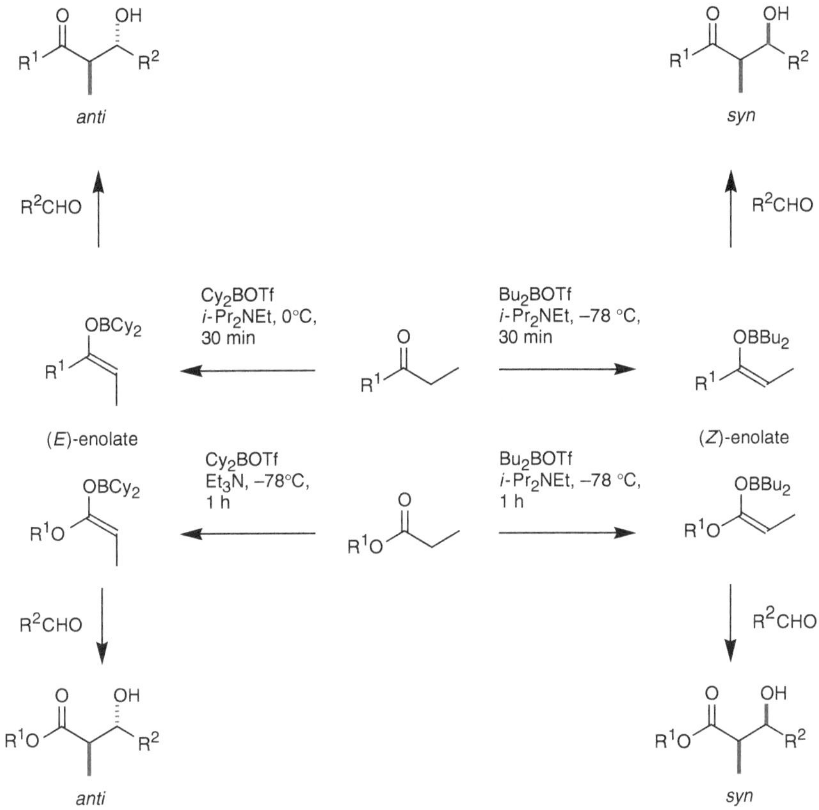

Scheme 2.3.3 Stereoselective formation of (*Z*)- or (*E*)-boron enolates of ketones or carboxylic esters and subsequent aldol addition

Reactions of boron enolates of chiral oxazolidinones with aldehydes represent one of the most popular applications of asymmetric boron enolate aldol reactions. Evans and coworkers were the first to describe stereoselective aldol additions mediated by chiral boron enolates of oxazolidinones.[34,35] As a consequence of (*Z*)-geometry of boron enolates employed in these reactions the formation of *syn*-configured products is observed. Extremely high selectivities were detected. When used with (*S*)-oxazolidinone **2** or (*R*)-oxazolidinone **4** an approach to both *syn*-configured enantiomers **3** and **5** is given (Scheme 2.3.4). In comparative reactions using the corresponding oxazolines **8** and **10** the enantioselectivities decrease (Eqs. 4 and 5, Scheme 2.3.4).

Substituents in α-position (R^1) are crucial for achieving these high selectivities.[37] No selectivities were observed in acetate aldol additions (R^1 = H). An explanation for the high enantioselectivities is given by considering transition states

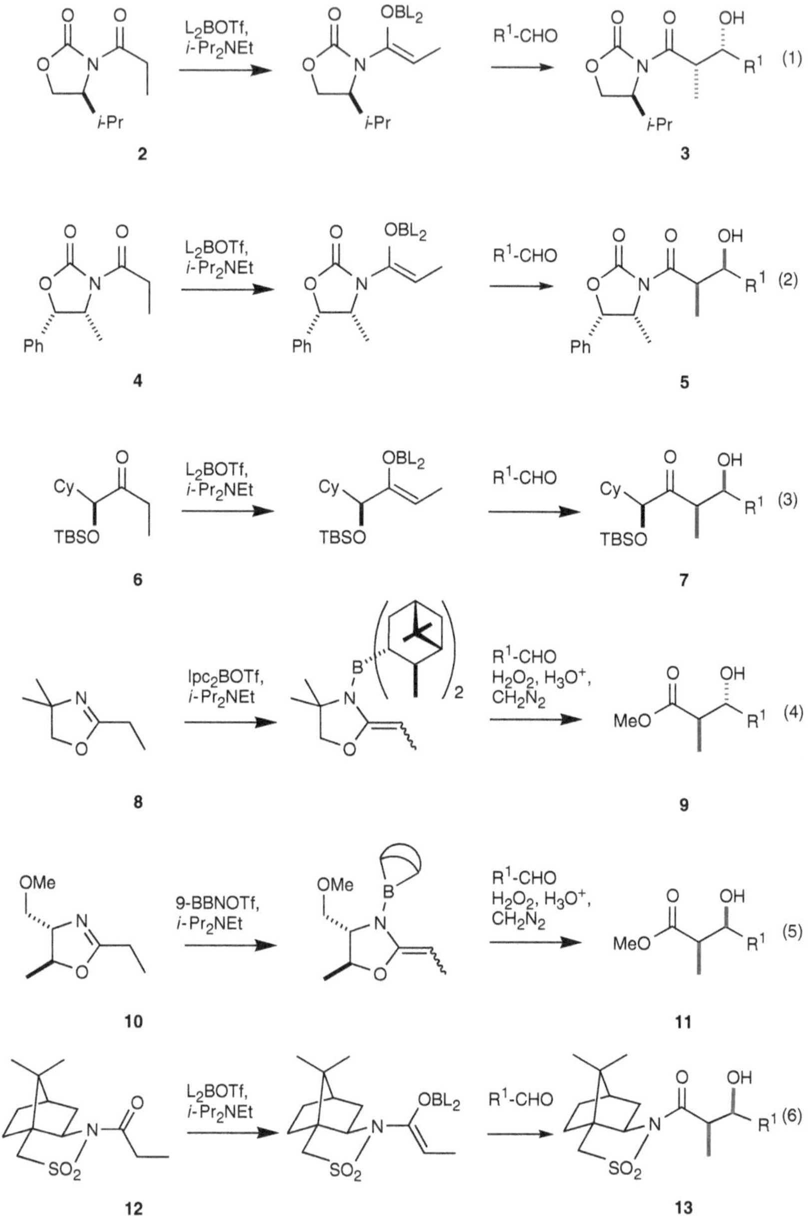

Scheme 2.3.4 Stereoselective aldol additions of boron enolates; Eqs. 1[26], 2[27], 3[28,29], 4[30], 5[31], 6[32,33]

in Scheme 2.3.5. The approach of the aldehyde (R^2) lies far away from the bulky control group favouring transition states **C** and **F**. Transition state models **C** and **F** can be differentiated by the orientation of the chiral centre towards the metal centre favouring the transition state **C** (allylic strain).

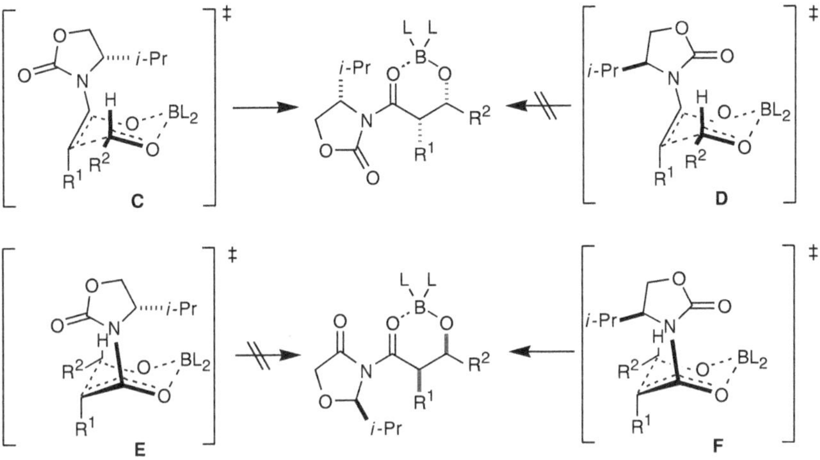

Scheme 2.3.5 Transition state models for the explanation of the *syn*-diastereoselectivity

In some cases the addition of Lewis acids changes the *syn*-selectivity to *anti*-selectivity (Schemes 2.3.6 and 2.3.7). This change in configuration can be explained at best by the proposed Lewis acid-mediated open transition state model **G** (Scheme 2.3.6).

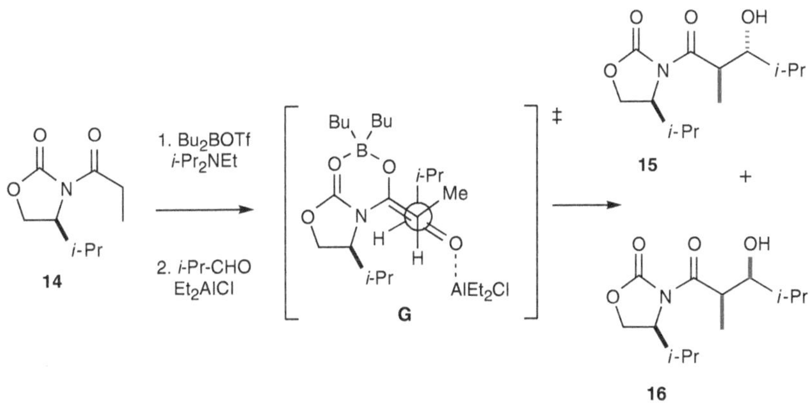

Scheme 2.3.6 *anti*-Selective aldol additions by addition of Et_2AlCl (**15**:**16**/95:5)

A more general route to chiral *anti*-configured aldol adducts via boron enolate chemistry is provided by the use of enantiomerically pure norephedrine esters **19**.[37] The stereochemical course of the aldol addition can be controlled by careful selection of the enolization reagent (Scheme 2.3.8).

For application of thioesters in this aldol methodology see Fanjul et al.[38] The discussed boron enolate aldol methodology has been successfully extended to the application of crotonate imides,[31,40] to isothiocyano derivates,[41] to chloroacetyl oxazolidinones[42] and to bromoacetyl oxazolidinones.[43] For an overview of boron enolate aldol reactions of carboxylic esters see Abiko.[44]

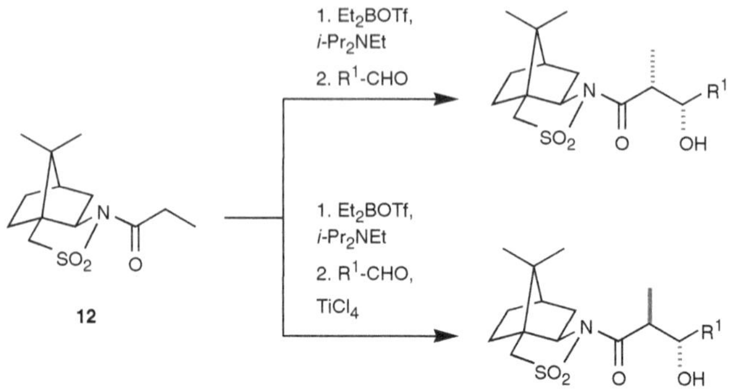

entry	R^1	equiv $TiCl_4$	temp (°C)	time (h)	yield (%)	**17 / 18**
1	Me	2	−78	2	73	4 / 96
2	Et	2	−78	2	77	2 / 98
3	*i*-Pr	2	−78	2	75	1 / 99
4	*t*-Bu	2	−40	15	73	10 / 90
5	Ph	1	−78	0.5	77	1 / 99

Scheme 2.3.7 *anti*-Selectivity by application of titanium(IV) chloride in aldol additions of boron enolates

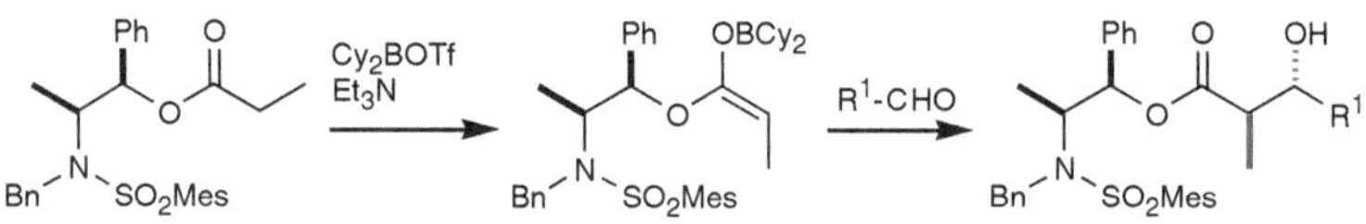

entry	R^1	yield (%)	*anti* / *syn*
1	Me	92	97 / 3
2	Et	90	96 / 4
3	*n*-Pr	95	95 / 5
4	*i*-Pr	95	98 / 2
5	Cy	91	95 / 5
6	*t*-Bu	96	>99 / 1
7	Ph	93	95 / 5
8	(*E*)-$CH_3CH = CH$	96	98 / 2
9	$H_2C = CCH_3$	97	96 / 4
10	$BnOCH_2CH_2$	94	95 / 5
11	$BnOCH_2C(CH_3)$	98	96 / 4

Scheme 2.3.8 *anti*-Selectivity by application of norephedrine esters

Tremendous comprehensive investigations on the stereochemical outcome of boron enolate aldol additions were carried out by Evans and coworkers. For an actual state of these investigations see Evans et al.[45] and the references cited in it.

Also, calculations of transition states to explain the observed stereoselectivities have been reported.[46] For investigations on the influence of the electronic nature of the substituents of boron enolates in aldol additions see Dias and others.[47] For attempts to realize a catalytic execution of boron enolate aldol additions see Mori and others.[48,49] These reactions were carried out even in water.[50] Asymmetric aldol additions using immobilized chiral boron enolates are described in Burke et al.[51]

The great reliability of boron enolate aldol additions is reflected by numerous applications in total syntheses of polyketides. For comprehensive reviews see Florence et al.[52] Some examples illustrate the power of this method in natural product total synthesis.

Paterson and coworkers used extensively their well-elaborated boron enolate aldol methodology in several total syntheses of polyketide natural products.[53] The first example represents a synthetic approach to discodermolide. Discodermolide – a marine polyketide – is a member of a group of natural products that act as microtubule-stabilizing agents and mitotic spindle poisons, which currently include well-known natural products such as paclitaxel, epothilones, laulimalide, peloruside and dictyostatin. Several different total syntheses were reported and they are discussed in Paterson and Florence.[54]

The construction of the requisite polypropionate arrays found in discodermolide was accomplished at very early stages by using several boron enolates in aldol additions (Scheme 2.3.9).[55] Boron-mediated aldol reaction of methyl ketone **22** and aldehyde **21** exploited remote 1,6-asymmetric induction from C-10 as indicated in transition state **H**. Enolization of **22** with Cy_2BCl/Et_3N and reaction with **21** gave the desired (5*S*)-configured discodermolide with a ratio of 95:5. In contrast, the analogous aldol reaction with the corresponding lithium enolate gave the (5*R*) adduct exclusively, as expected from Felkin-Anh control.

In a further study, Paterson and coworkers investigated the total synthesis of spongistatin – an antimitotic macrolide found in marine sponges. For this, they made extensive use of several boron enolate methods for the synthesis of the starting chiral compounds (Scheme 2.3.10).[56] This synthesis is one of the most deployments of boron enolate aldol methodologies in natural product synthesis.

A recent example of extensive deployment of Paterson's boron enolate methodology is the total synthesis of maurenone. Its strategy is shown in Scheme 2.3.11. Via a cascade of boron enolate aldol additions differently configured starting chiral aldehydes **51**, *ent*-**51**, **58** and *ent*-**58** (Scheme 2.3.12) and chiral ketones **52** and **61** (Scheme 2.3.13) were synthesized with a high degree of stereoselectivity.[57] Thus, an access to eight possible isomers of one enantiomeric series of maurenone is given. Comparison of the ^{13}C NMR data for the eight isomers with that reported for maurenone established the relative stereochemistry of the natural product.

Note that the simple change of the protecting groups (Bn $\rightarrow$ Bz and vice versa) and the application of different enolization methods provide a selective access

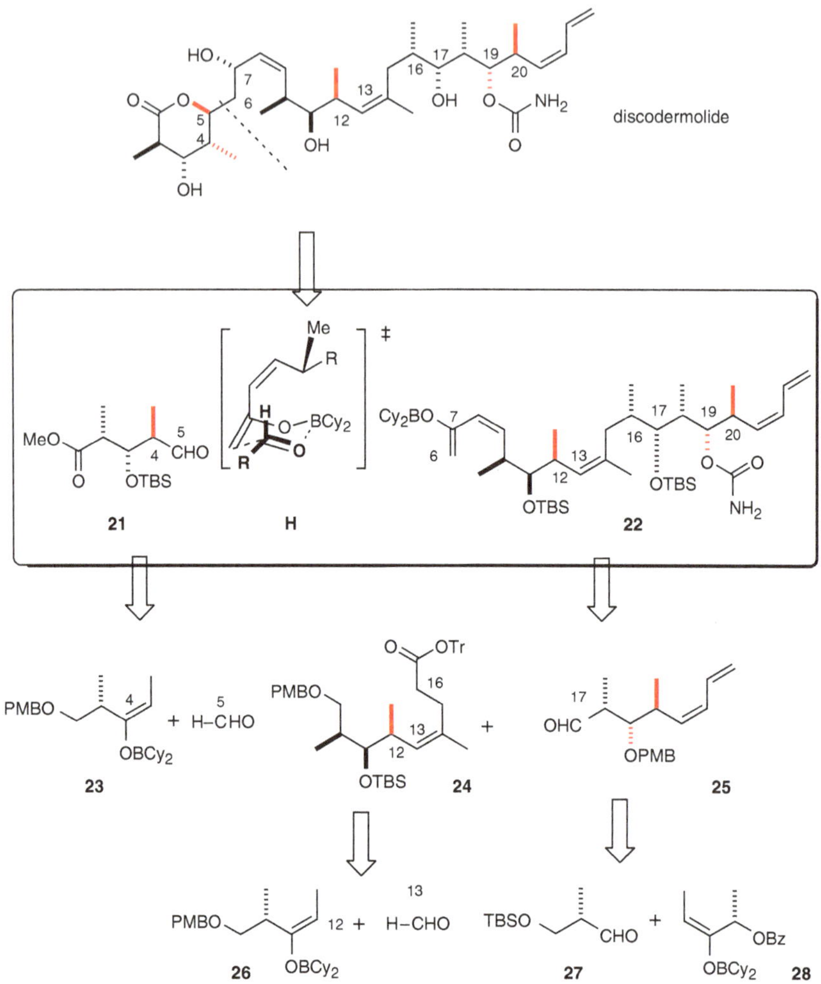

Scheme 2.3.9 Boron enolate aldol additions in total synthesis of discodermolide

to *syn-* and *anti-*configured aldehydes **51** and **58** with a high degree of diastereoselectivity (dr > 97/3). *Syn-*selectivity is obtained by strictly working at −78°C, whereas an *anti-*preference is observed by working from −78°C to 0°C.

For similar applications of boron enolates of dihydroxyacetone or erythrulose in aldol reactions to chiral α-branched aldehydes see Diaz-Oltra et al.[58]

Stereoselective boron enolate aldol additions were applied to total synthesis of elaiolide.[59] Elaiolide belongs to the efomycines – a class of anti-inflammatory agents isolated from microorganisms. The central stereopentad was obtained by *anti-*selective boron enolate aldol addition (Scheme 2.3.14). For an overview of different synthetic approaches to efomycines see Toshima et al.[60]

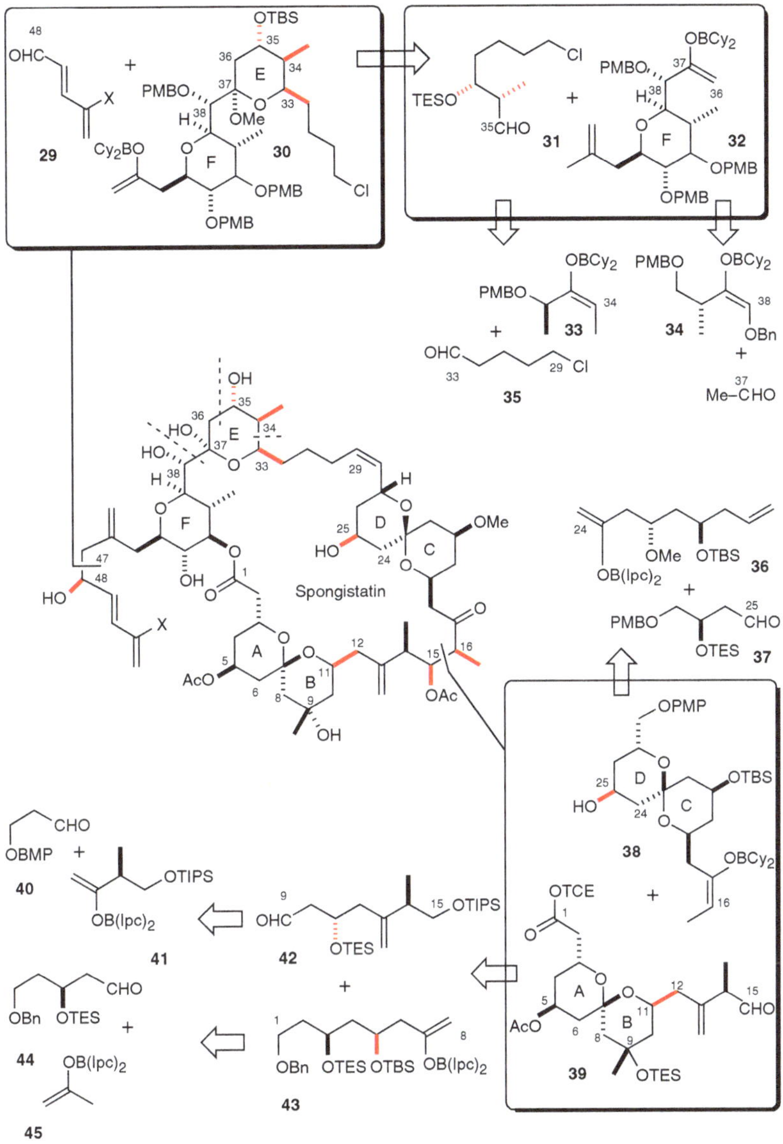

Scheme 2.3.10 Boron enolate aldol additions in total synthesis of spongistatin

An efficient synthesis of epothilone D analogs has been developed using highly stereoselective boron enolate aldol addition. The aldol adduct **67** was isolated in quantitative yields as a single stereoisomer (Scheme 2.3.15).[61]

During the total synthesis of australine a defined configured stereopentad was required. The construction of this stereopentad was accomplished by a boron enolate

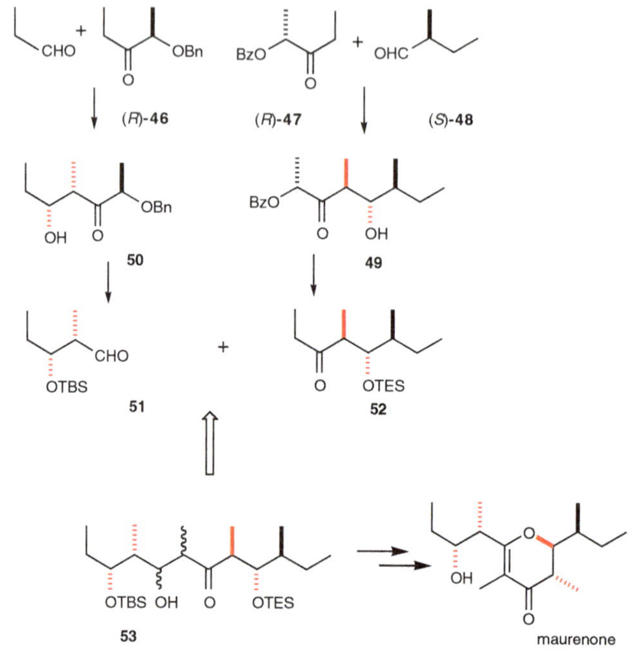

Scheme 2.3.11 Retrosynthetic strategy to maurenone

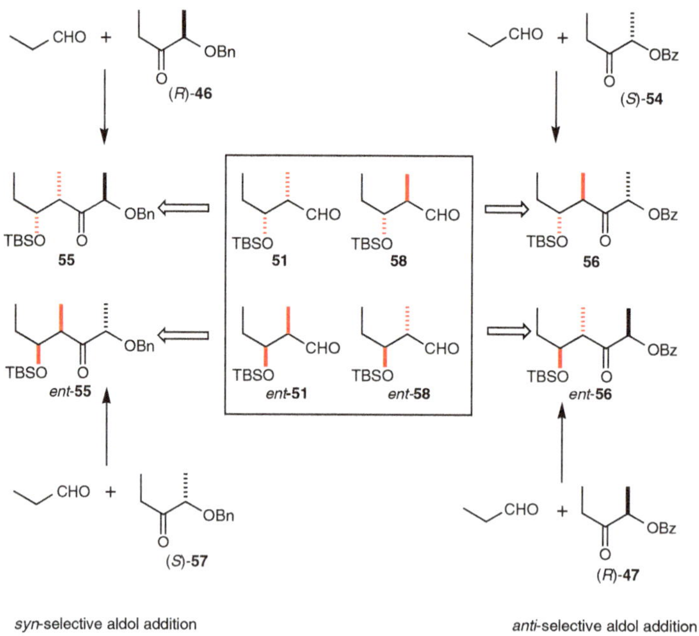

Scheme 2.3.12 Total synthesis of maurenone – stereoselective synthesis of starting aldehydes **51** and **58**

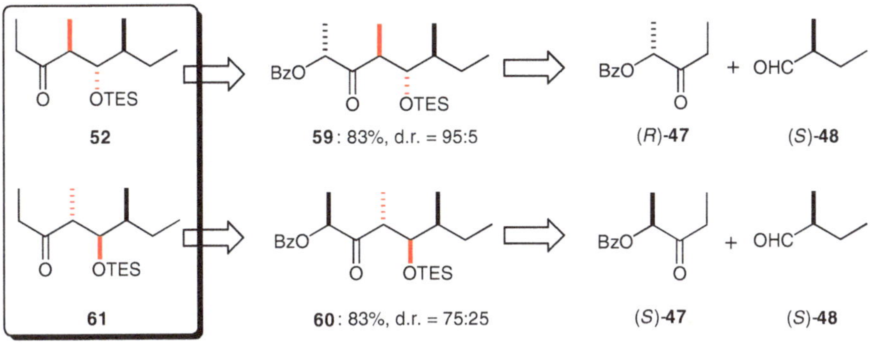

Scheme 2.3.13 Total synthesis of maurenone – stereoselective synthesis of starting *anti*-configured ketones **52** and **61**

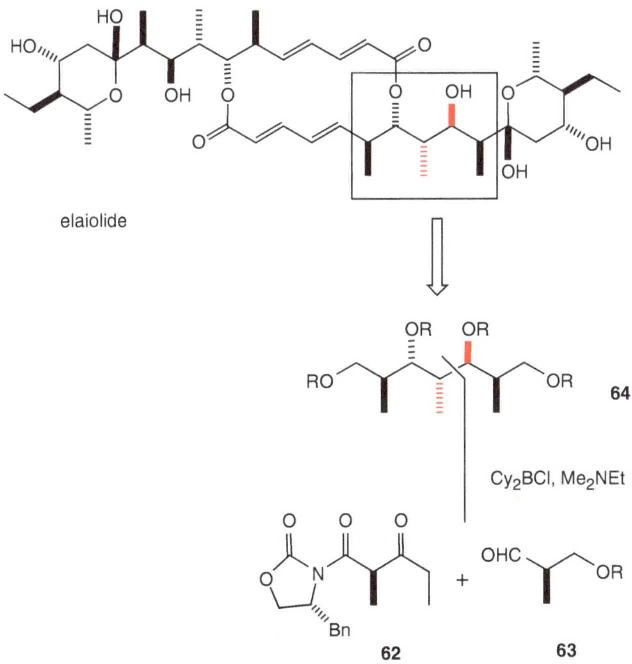

Scheme 2.3.14 Total synthesis of elaiolide

aldol step of chiral aldehyde **69** to chiral ketone **68** followed by $LiBH_4$-mediated reduction (Scheme 2.3.16).[62] For further results of stereocontrolled aldol additions of boron enolates of erythrulose derivates and chiral aldehydes see Marco et al.[63]

The α-oxygen-containing aldehyde **71** was deployed in a convergent total synthesis of pteridic acid A. Again, the Paterson group could demonstrate the

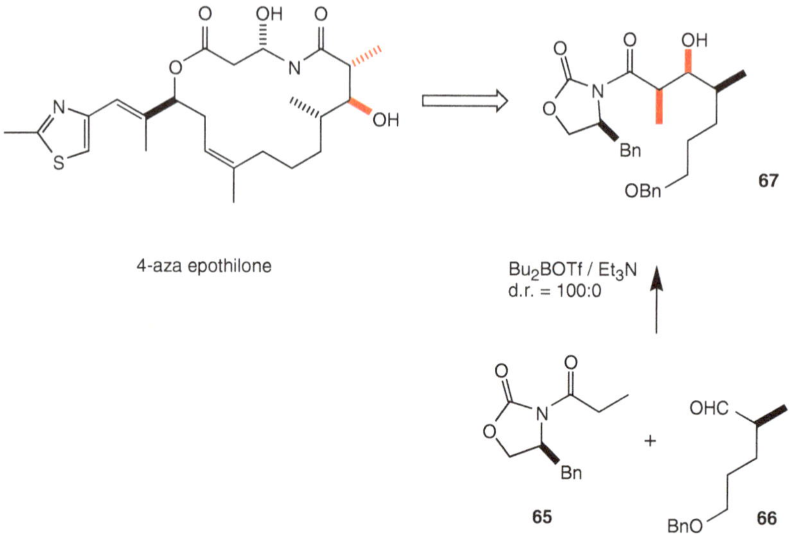

Scheme 2.3.15 Total synthesis of 4-aza epothilones

Scheme 2.3.16 Total synthesis of australine

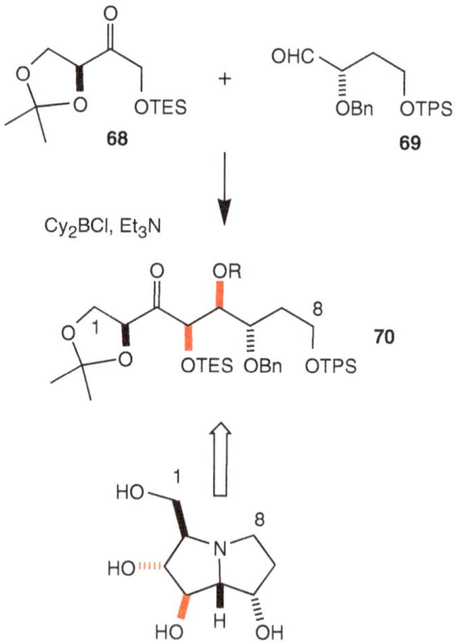

reliability of the boron enolate aldol methodology. The desired intermediates **73** and **75** were obtained with a high degree of *anti*-diastereoselectivity (Scheme 2.3.17).[64]

A rare case of 1.5-induction[65] during boron enolate aldol additions was investigated by Evans and coworkers. During the total synthesis of callipeltoside, the

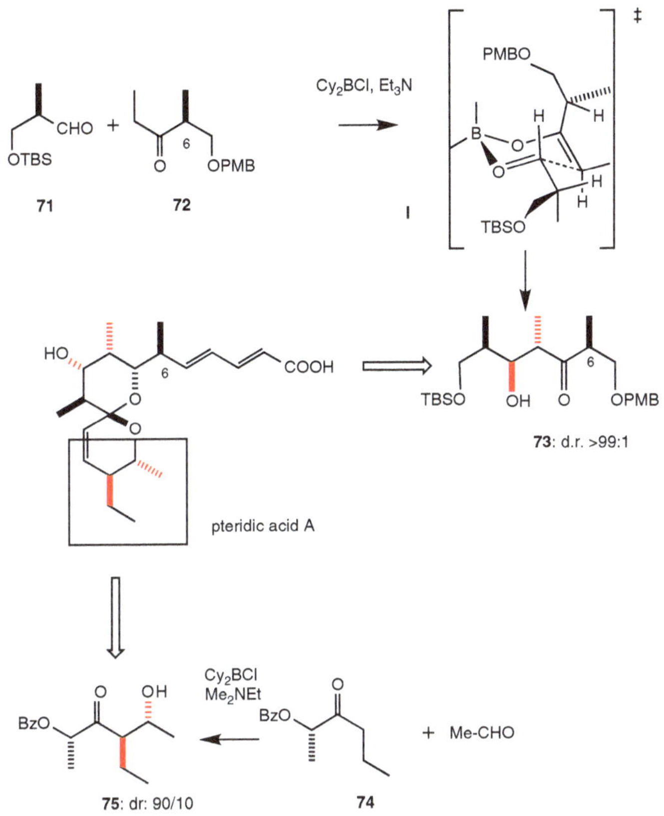

Scheme 2.3.17 Total synthesis of pteridic acid A

authors observed a matched/mismatched situation evoked by 1.5-induction in boron enolate aldol additions. In further investigations the authors could demonstrate that the stereochemical outcome of this aldol reaction depends on the configuration at C-5 of aldehydes **77** and **79**, whereas the protecting group at C-6 has no significant influence on the stereochemical outcome (Scheme 2.3.18).[66]

For recent investigations of asymmetric 1.2-induction in aldol additions of boron enolates to oxygen-substituted aldehydes see Evans et al.[67]

The initial stereoselective aldol step applying boron enolates of chiral benzyl-*N*-propionyl-2-oxazolidinone was used in several total syntheses of natural products, e.g. hexacyclinic acid.[68]

The C1–C11 subunit of tedanolide was constructed by a boron enolate aldol step. By the application of Bu_2BOTf in the presence of Et_3N this segment could be isolated with 35% yield favouring the desired *syn*-configuration (91:9) (Scheme 2.3.19).[69]

Further deployments of boron enolates were reported in total syntheses of rhizoxin D,[70] leucascandrolide,[71] apratoxin A (Oppolzer's sultam methodology),[72,73] sitophilur (Enders SAMP methodology).[74]

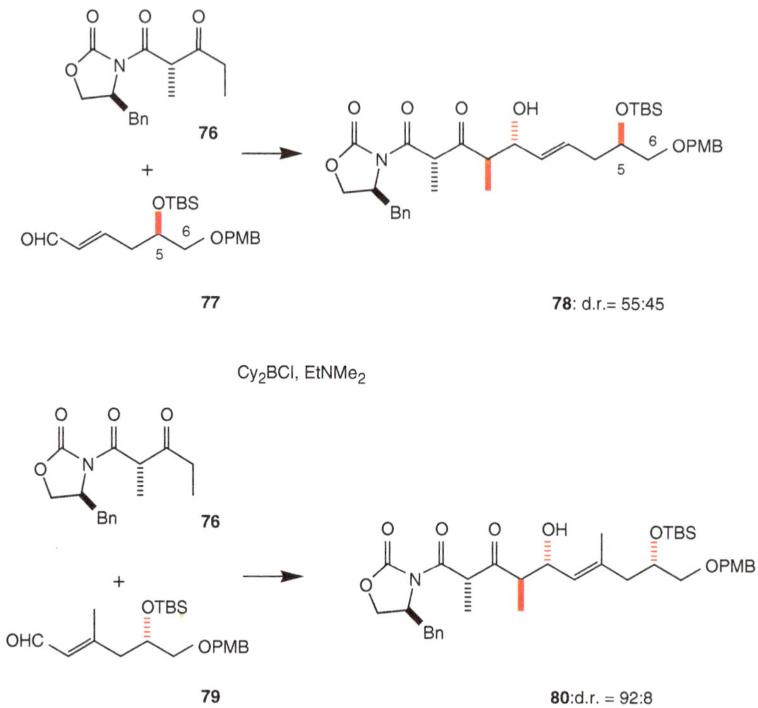

Scheme 2.3.18 1.5-Induction in boron enolate aldol additions

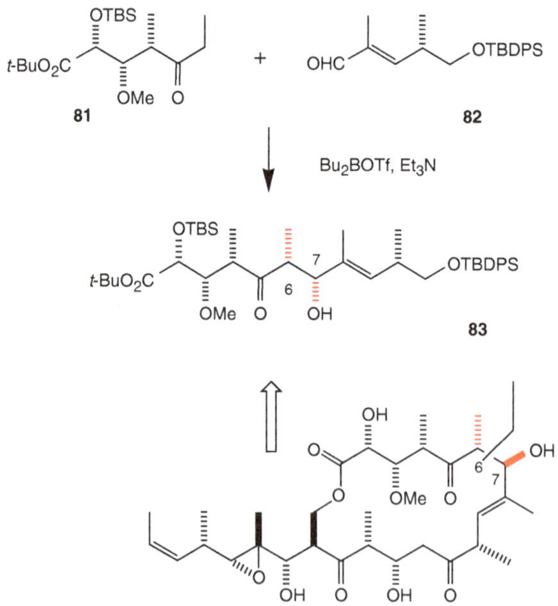

Scheme 2.3.19 Boron enolate aldol step in total synthesis of tedanolide

References

1. Mukaiyama T (1982) Org React 28:203
2. Cowden CJ, Paterson I (1997) Org Reactions 51:1
3. Kim BM, Williams SF, Masamune S (1993) In: Comprehensive Organic Synthesis, Trost BM, Fleming I, Heathcock CH (eds). Pergamon, Oxford, vol 2, p 239
4. Evans DA, Nelson JV, Taber TR (1982) Top Stereochem 13:1
5. Mukaiyama T, Matsuo JI (2004) In: Modern Aldol Reactions Mahrwald R (ed). Wiley-VCH, Weinheim, vol 1, p 127
6. Paterson I, Cowden CJ, Wallace DJ (2000) In: Modern Carbonyl Chemistry, Otera J (ed). Wiley-VCH, Weinheim, p 249
7. Mukaiyama T, Inoue T (1976) Chem Lett 559
8. Inoue T, Uchimaru Z, Mukaiyama T (1977) Chem Lett 153
9. Inoue T, Mukaiyama T (1980) Bull Chem Soc Jpn 53:174
10. Paterson I (1992) Pure Appl Chem 64:1821
11. Masamune S, Mori S, Van Horn DE, Brooks DW (1979) Tetrahedron Lett 20:1665
12. Van Horn DE, Masamune S (1979) Tetrahedron Lett 20:2229
13. Evans DA, Nelson JV, Vogel E, Taber TR (1981) J Am Chem Soc 103:3099
14. Evans DA, Vogel E, Nelson JV (1979) J Am Chem Soc 101:6120
15. Hirama M, Garvey DS, Lu LDL, Masamune S (1979) Tetrahedron Lett 20:3937
16. Abiko, A, Liu JF, Masamune S (1996) J Org Chem 61:2590
17. Abiko A (2002) Org Synth 79:116
18. Abiko A (2002) Org Synth 79:103
19. Brown HC, Ganesan K, Dhar R (1993) J Org Chem 58:147
20. Chow HF, Seebach D (1986) Helv Chim Acta 69:604
21. Hirama M, Masamune S (1979) Tetrahedron Lett 2225
22. Evans DA, Nelson JV, Vogel E, Taber TR (1981) J Am Chem Soc 103:3099
23. Hirama M, Garvey DS, Lu LDL, Masamune S (1979) Tetrahedron Lett 20:937
24. Cergol KM, Coster MJ (2007) Nature Protocols 2:2568
25. Abiko A, Liu JF, Masamune S (1996) J Org Chem 61:2590
26. Evans DA, Bartroli J (1982) Tetrahedron Lett 23:807
27. Evans DA, Bartroli J, Shih TL (1981) J Am Chem Soc 103:2127
28. Masamune S, Choy W, Kerdesky FAJ, Imperiali B (1981) J Am Chem Soc 103:1566
29. McCarthy PA (1982) Tetrahedron Lett 23:4199
30. Meyers AI, Yamamoto Y (1984) Tetrahedron 40:2309
31. Meyers AI, Yamamoto Y (1981) J Am Chem Soc 103:4278
32. Oppolzer W, Blagg J, Rodriguez I, Walther E (1980) J Am Chem Soc 112:2762
33. Oppolzer W (1988) Pure Appl Chem 60:39
34. Evans DA, Takacs JM, McGee LR, Ennis MD, Mathre DJ, Bartroli J (1981) Pure Appl Chem 53:1109
35. Evans DA (1982) Aldrichimica Acta 15:23
36. Braun M (1981) Angew Chem Int Ed Engl 26:24
37. Inoue T, Liu JF, Buske DC, Abiko A (2002) J Org Chem 67:5250
38. Fanjul S, Hulme AN, White JW (2006) Org Lett 8:4219
39. Evans DA, Sjorgren EB, Bartroli J, Dow RL (1986) Tetrahedron Lett 27:4957
40. Evans DA, Dow RL, Shih TL, Takacs JM, Zahler R (1990) J Am Chem Soc 112:5290
41. Evans DA, Weber AE (1986) J Am Chem Soc 108:6557
42. Evans DA, Sjorgren EB, Bartroli J, Dow RL (1986) Tetrahedron Lett 27:4957
43. Evans DA, Sjorgren EB (1985) Tetrahedron Lett 26:3783
44. Abiko A (2004) Acc Chem Res 37:387
45. Evans DA, Cee V, Siska SJ (2006) J Am Chem Soc 128:9433
46. (a) Goodman JM, Paton RS (2007) Chem Commun 2124; (b) Paton RS, Goodman JM (2008) J Org Chem 73:1253; (c) Paton, RS, Goodman JM (2006) Org Lett 8:4299

47. (a) Dias LC, Aguilar AM (2006) Org Lett 8:4629; (b) Dias LC, De Marchi AA, Ferreira MAB, Aguilar AM (2007) Org Lett 9:4869; (c) Dias LC, Aguilar AM (2008) Chem Soc Rev 37:451
48. Mori Y, Kobayashi J, Manabe K, Kobayashi S (2002) Tetrahedron 58:8263
49. Mori Y, Manabe K, Kobayashi S (2001) Angew Chem Int Ed Engl 40:2815
50. Aelvoet K, Batsanov AS, Blatch AJ, Grosjean C, Patrick LGF, Smethurst CA, Whiting A (2008) Angew Chem Int Ed 47:768
51. Burke MD, Berger EM, Schreiber SL (2003) Science 302:613
52. (a) Florence GJ, Gardner NM, Paterson I (2008) Nat Prod Rep 25:342; (b) Yeung KS, Paterson I (2005) Chem Rev 105:4237; (c) Norcross RD, Paterson I (1995) Chem Rev 95:2041
53. (a) Paterson I, Wallace DC, Cowden CD (1998) Synthesis 639; (b) Paterson I, Norcross RD, Ward RA, Romea P, Lister MA (1994) J Am Chem Soc 116:11287
54. Paterson I, Florence GJ (2003) Eur J Org Chem 2193
55. (a) Paterson I, Wren SP (1993) Chem Commun 1790; (b) Paterson I, Schlapbach A (1995) Synlett 498
56. (a) Paterson I, Chen DYK, Coster MJ, Acena JL, Bach J, Gibson KR, Keown LE, Oballa RM, Trieselmann A, Wallace DJ, Hodgson AP, Norcross RD (2001) Angew Chem Int Ed Engl 40:4055; (b) Paterson I, Coster MJ, Chen DYK, Gibson KR, Wallace DJ (2005) Org Biomol Chem 3:2410; (c) Paterson I, Coster MJ, Chen DYK, Oballa RM, Wallace DJ, Norcross RD (2005) Org Biomol Chem 3:2399
57. Crossman JS, Perkins MV (2006) J Org Chem 71:117
58. Diaz-Oltra S, Carda M, Murga J, Falomir E, Marco JA (2008) Chem Eur J 14:9240
59. Barth R, Mulzer J (2008) Tetrahedron 64:4718; for further total synthesis see also (a) Evans DA, Fitch DM (1997) J Org Chem 62:454; (b) Paterson I, Lombart HG, Allerton C (1999) Org Lett 1:19; (c) Paterson I, Man J (1997) Tetrahedron Lett 38:695
60. (a) Toshima K, Tatsuta K. Kinoshita M (1986) Tetrahedron Lett 27:4741; (b) Toshima K, Tatsuta K, Kinoshita M (1988) Bull Chem Soc Jpn 61:2369; (c) Seebach D, Chow HF, Jackson RFW, Lawson K, Sutter MA, Thaisrivongs S, Zimmermann J (1985) J Am Chem Soc 107:5292; (d) Evans DA, Fitch DM (1997) J Org Chem 62:454; (e) Paterson I, Lombart HG, Allerton C (1999) Org Lett 1:19; (f) Paterson I, Man J (1997) Tetrahedron Lett 38:695
61. Cachoux F, Schaal F, Teichert A, Wagner T, Altmann KH (2004) Synlett 2709
62. Ribes C, Falomir E, Carda M, Marco JA (2007) Org Lett 9:77
63. Marco JA, Carda M, Díaz-Oltra S, Murga J, Falomir E, Roeper H (2003) J Org Chem 68:8577
64. Paterson I, Anderson EA, Findlay AD, Knappy CS (2008) Tetrahedron 64:4768
65. Mahrwald R (1999) Chem Rev 99:1095
66. Evans DA, Burch JD, Hu E, Jaeschke G (2008) Tetrahedron 64:4671
67. Evans DA, Siska SJ, Cee VJ (2003) Angew Chem Int Ed 42:1761
68. Evans DA, Starr JT (2004) Angew Chem Int Ed 41:1787
69. Loh TP, Feng LC (2001) Tetrahedron Lett 42:6001
70. Jiang Y, Hong J, Burke SD (2004) Org Lett 6:1445
71. (a) Paterson I, Tudge M (2003) Angew Chem Int Ed 42:343; (b) Paterson I, Tudge M (2003) Tetrahedron 59:6833
72. Zou B, Wie J, Cai G, Ma D (2003) Org Lett 5:3503
73. Chen J, Forsyth CJF (2003) J Am Chem Soc 125:8734
74. (a) Enders D, Lohray BB (1988) Angew Chem 100:594; (b) Enders D, Lohray BB (1987) Angew Chem 99:395; (c) Enders D, Prokopenko OF, Raabe G, Runsink J (1996) Synthesis 1095

2.4 Titanium Enolates

Titanium enolate-based aldol additions have a tremendous synthetic potential. Titanium reagents are readily available, inexpensive, nontoxic and easy to handle.[1,2] First aldol additions of aldehydes to titanium enolates were described by Reetz and coworkers.[3] The titanium enolates were generated mostly by transmetallation of corresponding lithium enolates with $ClTi(Oi\text{-}Pr)_3$ or $ClTi(NR_2)_3$. Titanium enolates were found to add *syn*-selectively to aldehydes irrespective of the geometry of the enolates (Scheme 2.4.1).

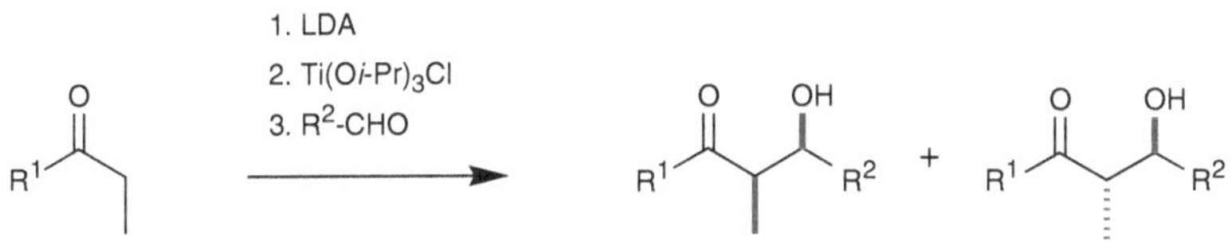

entry	R^1	R^2	*syn* / *anti*
1	Et	Ph	89 / 11
2	Et	*t*-Bu	81 / 19
3	*t*-Bu	Ph	87 / 13
4	Ph	Ph	87 / 13

Scheme 2.4.1 Aldol additions to titanium enolates with aldehydes

Harrison reported an aldol process where the titanium enolates were generated in situ by applying $TiCl_4$ and Et_3N.[4] Evans and coworkers used a combination of $TiCl_4$ and $i\text{-}Pr_2NEt$ for *syn*-selective aldol addition. The described method is more efficient and can be used for many different kinds of substrates (Scheme 2.4.2).[5] For results with (*S*)-2-benzyloxy-3-pentanone in aldol additions under these conditions see Rodriguez-Cisterna et al. and Pellicena et al.[6,7]

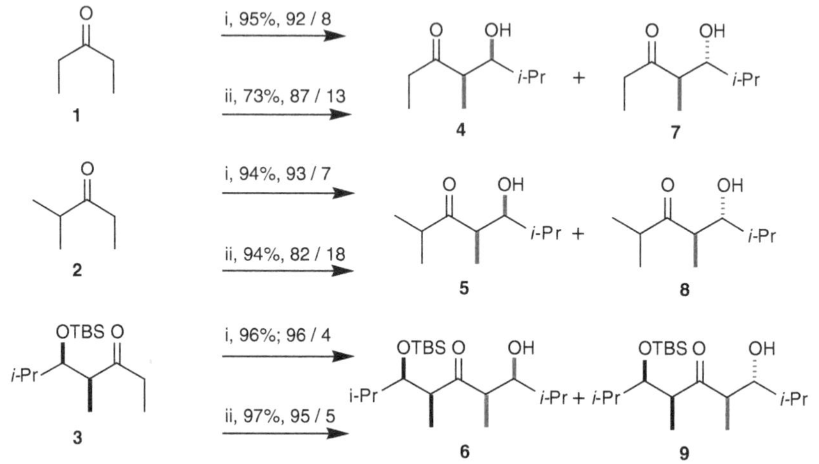

Scheme 2.4.2 *syn*-Selective aldol additions of isobutyraldehyde by employment of different reaction conditions: (i) $TiCl_4$, Pr_2NEt, *i*-Pr-CHO; and (ii) $TiCl_4$, Et_3N, *i*-Pr-CHO

Also, *syn*-selective aldol additions were obtained by using the same method with titanium enolates of thioesters (Scheme 2.4.3).[8]

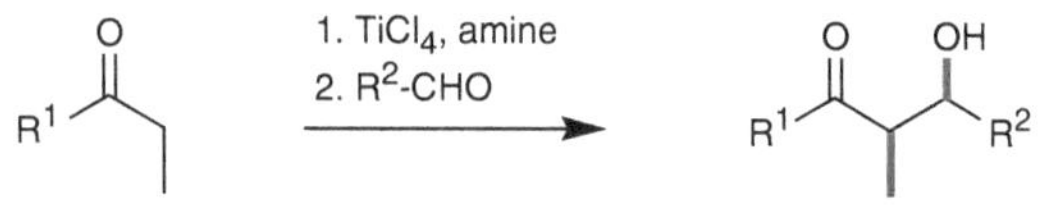

entry	R^1	R^2	yield (%)	*syn* / *anti*
1	PhS	Ph	70	85 / 15
2	PhS	*n*-Pr	68	78 / 22
3	PhS	*i*-Pr	65	89 / 11
4	*t*-BuS	Ph	77	86 / 14
5	*t*-BuS	*n*-Pr	72	69 / 31
6	*t*-BuS	*i*-Pr	75	71 / 29
7	2-$MeOC_6H_4O$	Ph	71	87 / 13
8	C_6F_5S	Ph	40	95 / 5

Scheme 2.4.3 Range of aldol additions of thioesters mediated by titanium(IV) chloride and amines

For a comparison of this method with boron and tin enolate aldol additions see Solsona et al.[9] Application of this protocol was found in the total synthesis of 2-epibotcinolide.[10]

Only a few examples of *anti*-selective titanium-mediated aldol additions have been published. By using cyclopentadienyl titanium enolates of *N*-propionylpyrrolidine *anti*-configured aldol adducts were obtained in good to high stereoselectivities[11] Aldol additions of α-benzyloxythioester-derived chlorotitanium enolates to aldehydes provide aldol adducts with a high degree of *anti*-diastereoselectivity.[12] Kazmaier and coworker reported the synthesis of *anti*-configured α-amino aldol adducts.[13] Aldol additions of enolizable aldehydes with diethylketone in the presence of titanate complexes provided *anti*-configured products with a high degree of stereoselectivity (Scheme 2.4.4).[14]

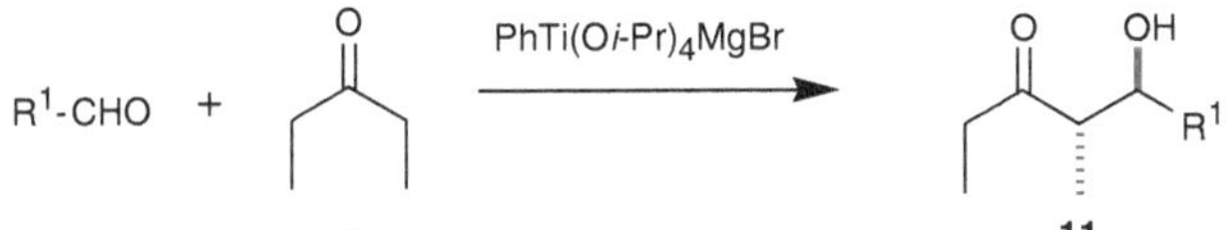

entry	R^1	yield (%)	*syn* / *anti*
1	*n*-Pr	81	16 / 84
2	*i*-Pr	72	2 / 98
3	*t*-Bu	78	2 / 98
4	Cy	77	2 / 98

Scheme 2.4.4 *anti*-Selective aldol additions in the presence of titanate complexes

Trisubstituted titanium enolates were reacted with aldehydes under kinetic conditions to give quantitatively *anti*-configured aldol adducts.[15] These results are consistent with those obtained in the lithium enolate series. Results obtained by using different titanium enolates of 2-methylcyclohexanone **12** in aldol additions are described in Scheme 2.4.5.

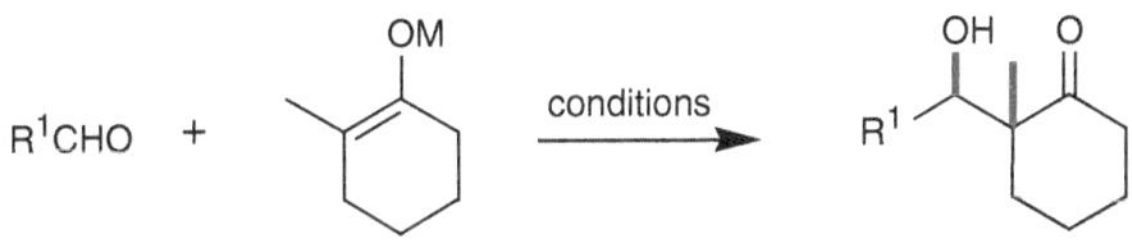

entry	M	R	conditions	*syn* / *anti*	yield (%)
1	$Ti(Oi\text{-}Pr)_3$	Ph	hexane, –72 °C, 1 h	5 / 95	94
2	$TiCl_3$	Ph	$CHCl_2$, –72 °C, 5 s	12 / 88	29
3	$TiCl_3$	*n*-Pr	$CHCl_2$, –72 °C, 5 min	9 / 91	56

Scheme 2.4.5 Aldol additions of trisubstituted titanium enolates

A further example of deployment of trisubstituted titanium enolates is the total synthesis of zaragozic acid. Evans and coworkers used aldol addition of titanium enolates in their synthetic approach to zaragozic acid.[16] The starting trimethyl silyl enolether **14** was generated from tartaric acid, which was then transformed into the corresponding titanium enolate (Scheme 2.4.6). The latter could be reacted with

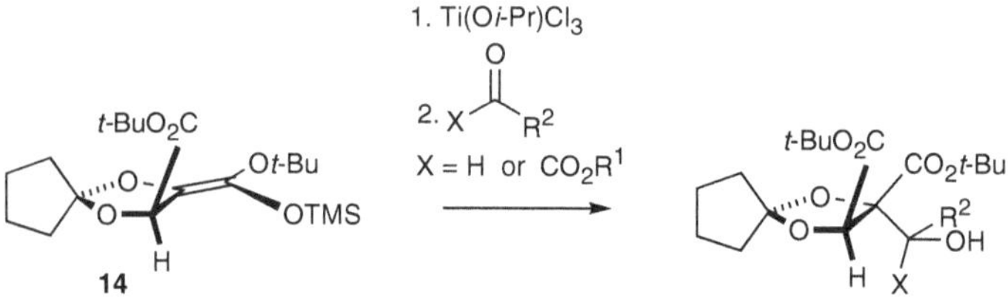

Entry	carbonyl compound	aldol adduct	selectivity	yield (%)
1	Bu-CHO	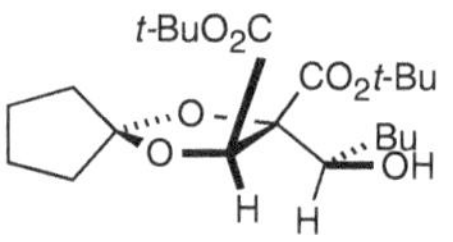	91 / 9	48
2	Ph-CHO	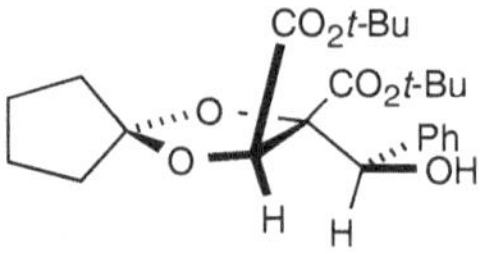	88 / 12	67
3	$BnOCH_2$-CHO	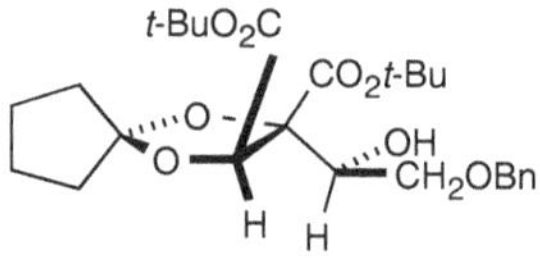	91 / 9	64
4	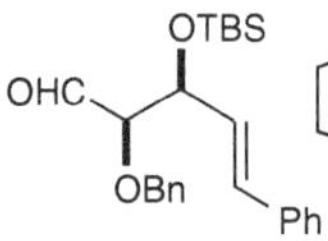	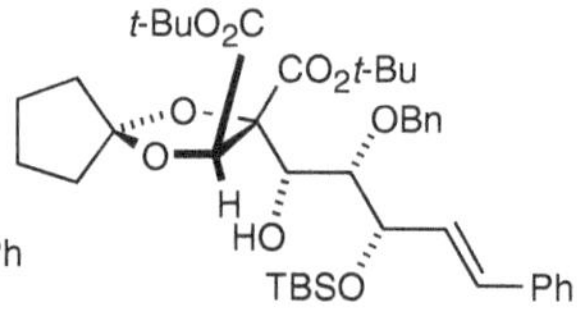	95 / 5	80
5	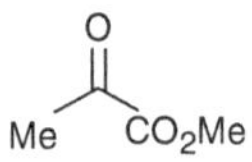	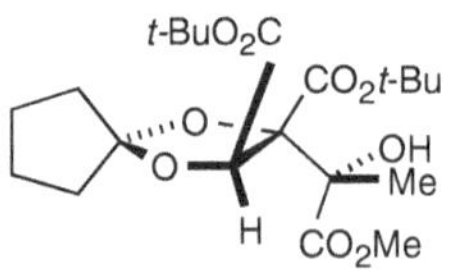	95 / 5	66
6	MeO_2C-CHO	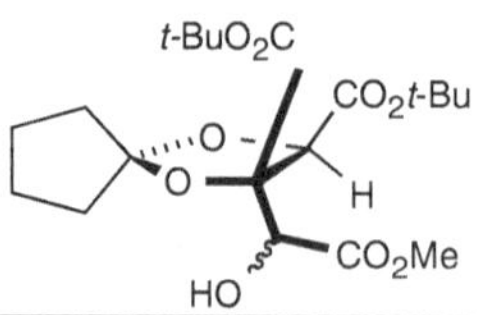	50 / 50	44

Scheme 2.4.6 Aldol additions to titanium enolates of (2S, 3S)-tartaric acid derivatives

several carbonyl compounds (aldehydes and ketoesters). The systematic results reported in Scheme 2.4.6 may be rationalized by invoking the proposed transition state model (Fig. 2.4.1). The chair-like transition state **A** orients the α-ketoester on the face of the tartrate enolate opposite to the *tert*-butyl ester. Thus, the methyl ester of the electrophile occupies a pseudo-axial orientation allowing chelation to titanium. The stereoinduction observed for unfunctionalized aldehydes (entries 1 and 2) can be rationalized by a pseudo-equatorial orientation of alkyl or aryl groups in the closed transition state model **B**. On the other hand, benzyloxyacetaldehyde realizes its chelate potential via pseudo-axial orientation of the benzyl ether (transition state model **C**). An identical arrangement leads to the selectivity observed with a more highly functionalized aldehyde (entry 4). Perhaps the highly activated nature of the aldehyde in entry 6 leads to a less rigid transition state and thus loss of stereochemical induction.

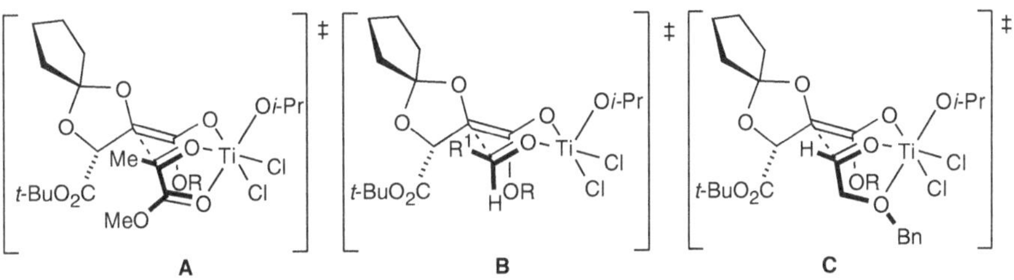

Fig. 2.4.1 Proposed transition states given for adol additions of (2*R*, 3*R*)-tartaric acid derivatives

To obtain optically active aldol adducts of titanium enolates amino acid-derived chiral auxiliaries were first tested by Thornton and coworkers.[17,18,19] The procedure involved transmetallation of lithium enolates with $ClTi(Oi\text{-}Pr)_3$. By varying the amount of titanium, the ratio of products **17** and **18** could be varied as shown in Scheme 2.4.7. Complications that arose from solubility problems were later solved by Evans et al.[20] Direct generation of titanium enolates in dichloromethane with amine bases and subsequent aldol addition yielded aldol adducts with a high degree of *syn*-stereoselectivity (Scheme 2.4.2).

Subsequently, Thornton and Bonner developed camphor-derived chiral auxiliaries. Moderate to good *syn*-selectivities were observed in these aldol reactions.[21] An improvement of *syn*-selectivity was achieved by the use of an oxazinone derivative of camphor in these reactions.[22] High *syn*-selectivities of chelation-controlled products were obtained. Yan and coworkers reported an extension of this method.[23,24] Application of titanium enolates of *N*-tosylnorephedrine,[25] chiral stilbenediamine,[26] and tetrahydro-1.3.4-oxadiazin-(2)-one[27] in aldol reactions were described.

At this time Gosh and coworkers described the first application of aminoindanol-derived asymmetric aldol reactions.[28] Chiral propionic acid esters of optically active aminoindanol **23** were transformed in situ to the corresponding titanium enolates with $TiCl_4$ in the presence of bases. The configurative outcome of this reaction depends on the substitution pattern of aldehydes used (Scheme 2.4.8). Oxygen-containing aldehydes yield aldol adducts with excellent degrees of

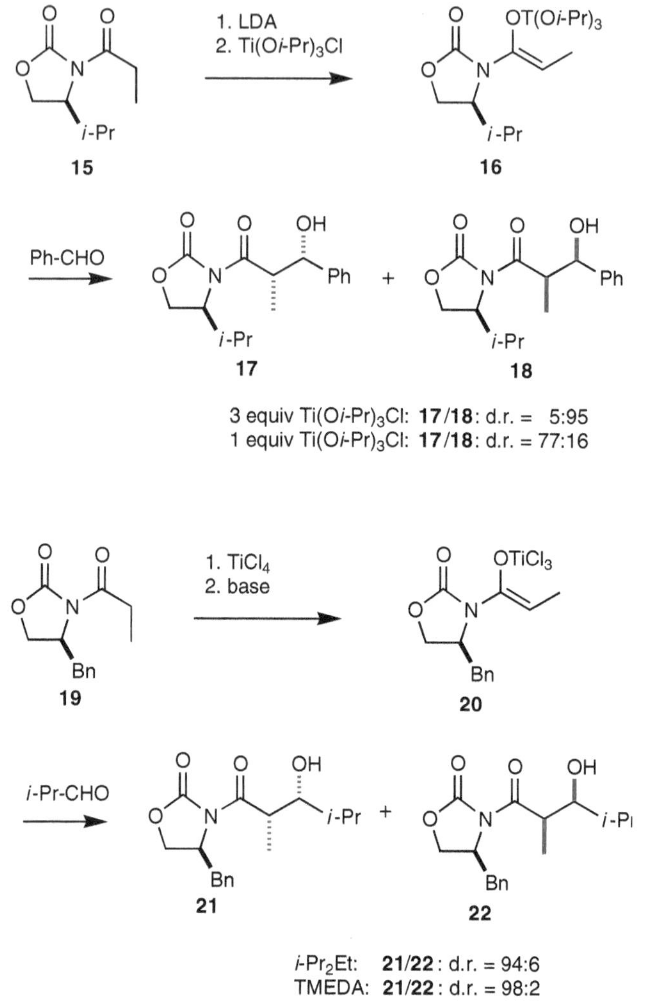

Scheme 2.4.7 *syn*-Selective aldol reactions of oxazolidine chiral auxiliaries

syn-diastereoselectivities.[29] This resulted from a chelation control as depicted in transition state model **D** (Scheme 2.4.8). When used with monodentate aldehydes *anti*-configured aldol adducts **25** were obtained.

Further investigations revealed that even aldol additions of corresponding simple amino alcohol-derived chiral auxiliaries result in excellent degrees of *syn*-diastereoselectivities.[30]

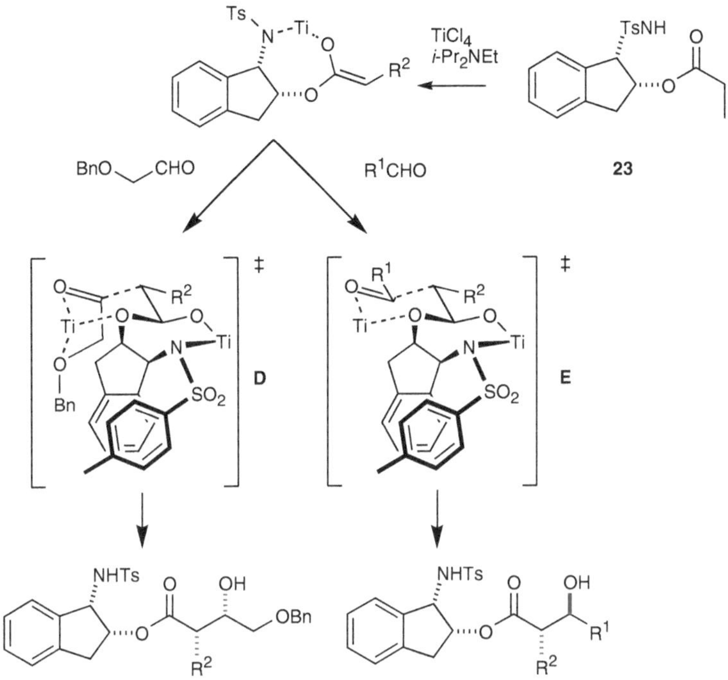

entry	R^1	R^2	yield (%)	*syn* / *anti*
1	$BnOCH_2$	Me	84	98 / 2
2	$BnO(CH_2)_2$	Me	51	98 / 2
3	$BnOCH_2$	Bn	84	99 / 1
4	$BnO(CH_2)_2$	Bn	51	99 / 1
5	$BnO(CH_2)_3$	Bn	55	94 / 6
6	$BnOCH_2$	*i*-Bu	83	99 / 1
7	$BnO(CH_2)_2$	*i*-Bu	56	99 / 1
8	*i*-Bu	Me	92	1 / 99
9	*i*-Bu	Bn	91	1 / 99
10	*i*-Bu	*i*-Bu	83	1 / 99

Scheme 2.4.8 Aminoindanol-derived asymmetric aldol additions

By deployment of (1*R*, 2*S*)-configured aminoindanole **26** in the same protocol *anti*-configured aldol products **27** were isolated with a high degree of diastereoselectivity (Schemes 2.4.9 and 2.4.10).[31]

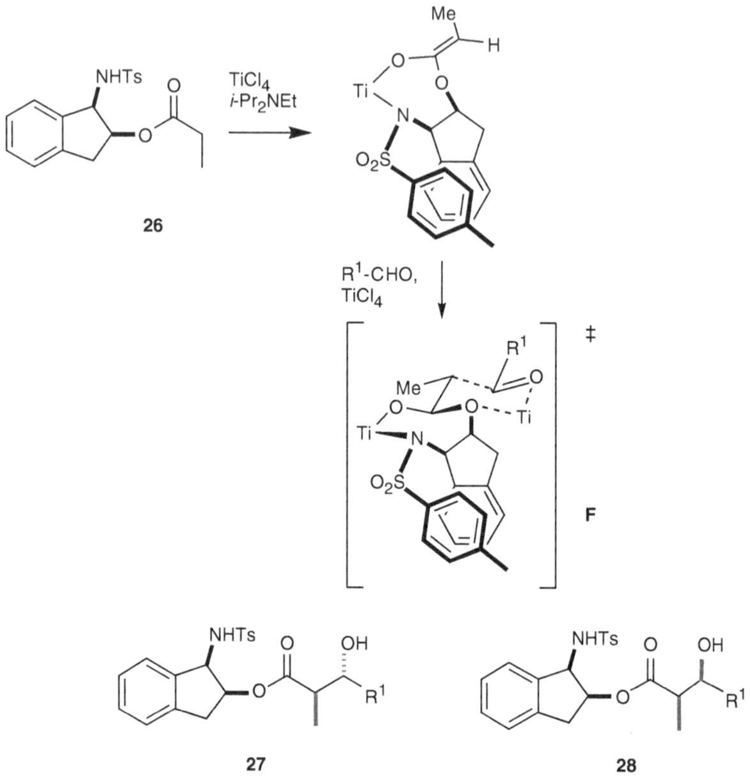

entry	R^1	yield (%)	*anti* / *syn*
1	*i*-Bu	97	>99 / 1
2	Me	50	85 / 15
3	*n*-Pr	74	95 / 5
4	*i*-Pr	91	85 / 15
5	MeCH=CH	41	95 / 5
6	PhCH=CH	63	99 / 1
7	Ph	85	45 / 55

Scheme 2.4.9 *anti*-Selective asymmetric aldol additions

Applications of this methodology were reported in total synthesis of cryptophycin B and arenastatin A,[32] brecanavir or darunavir.[33]

Crimmins and coworkers developed oxazolidinethione[34] and thiazolidinethione[35] as chiral auxiliaries to demonstrate their utility in titanium enolate aldol additions.

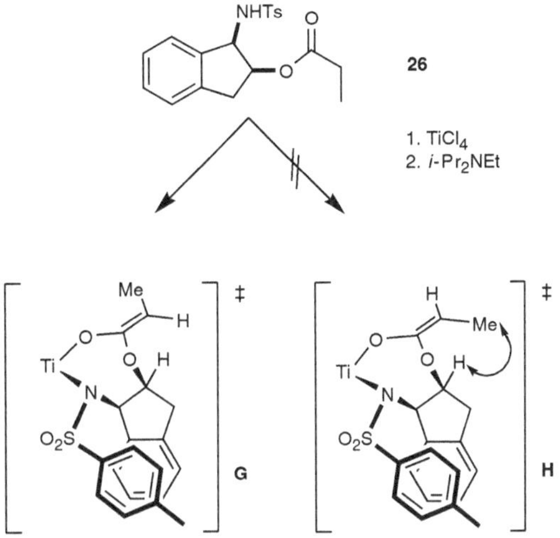

Scheme 2.4.10 Transition state models for the explanation of the *anti*-configuration

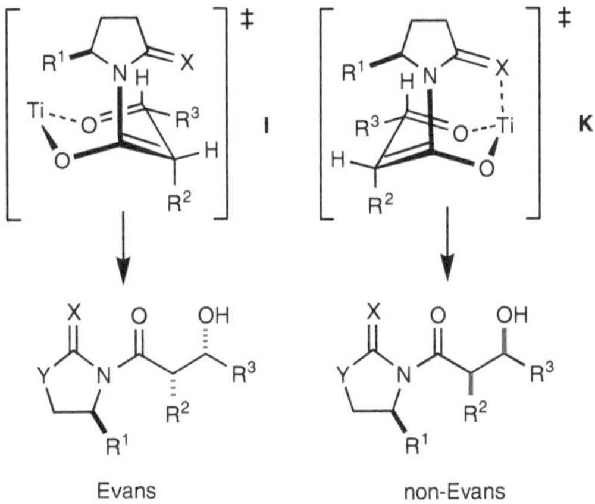

Scheme 2.4.11 Transition state models for *syn*-configured 'non-Evans' and 'Evans' aldol products

The reaction proceeds via the six-membered Zimmermann–Traxler transition state. In contrast to corresponding reactions in the boron enolate series an additional chelation to titanium is therefore possible (Scheme 2.4.11). Thus, an approach to the 'non-Evans' *syn*-aldol product is given. The optional use of amounts of amine bases and $TiCl_4$ yield the 'Evans' **29** or the 'non-Evans' **28** *syn*-configured aldol product with excellent diastereoselectivities (Scheme 2.4.12).[36]

Sparteine was found to be the amine of choice. It has no influence on the asymmetric induction but has a dramatic rate enhancement effect on the reactions.

Recently, an efficient strategy for the synthesis of *syn*- and *anti*-aldol adducts was published. By simply inverting the addition sequence of base and aldehydes an optional access to different configured aldol adducts is given (Scheme 2.4.13).[37]

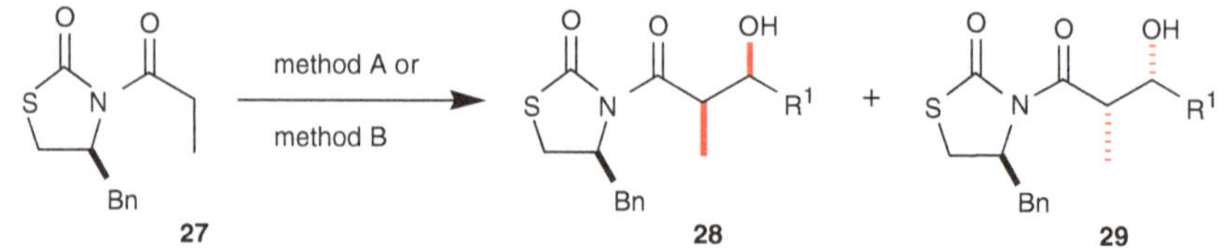

Method A: 1. 1.0 equiv. $TiCl_4$, 2. 2.5 equiv (–)-sparteine, 3. R^1-CHO
Method B: 1. 2.0 equiv. $TiCl_4$, 2. 1.1 equiv *i*-Pr_2NEt, 3. R^1-CHO

entry	method	R^1	yield (%)	**29** / **28** / *anti*
1	A	*i*-Pr	70	99 / 1 / 0
2	A	Ph	89	97 / 2 / 1
3	A	MeCH = CH	65	97 / 2 / 1
4	B	*i*-Pr	87	0 / 95 / 5
5	B	Ph	88	1 / 98 / 1
6	B	MeCH = CH	81	0 / 95 / 5

Scheme 2.4.12 Oxazolidinethione-based *syn*-selective aldol additions

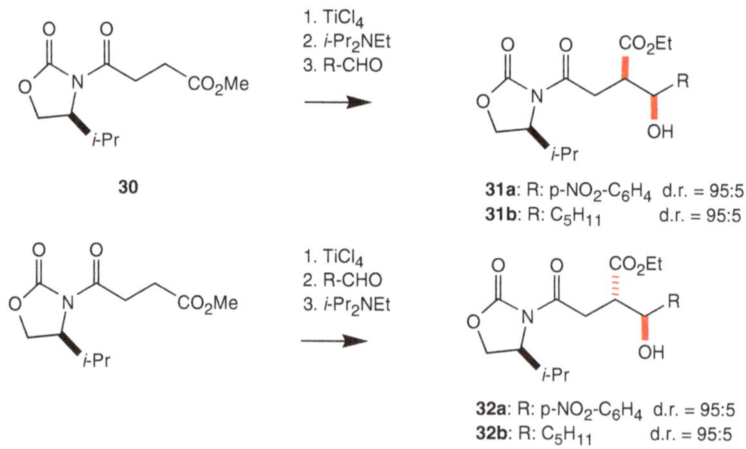

Scheme 2.4.13 Optional deployment of conventional and inverse addition sequence of bases and aldehydes

This high *syn*-stereoselectivity is observed even in aldol reactions of chiral α-substituted enolates.[38,39,40] Evans and coworkers intensively studied these reactions to install the desired configuration in polyketide natural products (Scheme 2.4.14).[41,42] When $Sn(OTf)_2$ is applied instead of $TiCl_4$ a reversal of selectivity is observed and compound **35** is the main product (Scheme 2.4.14).

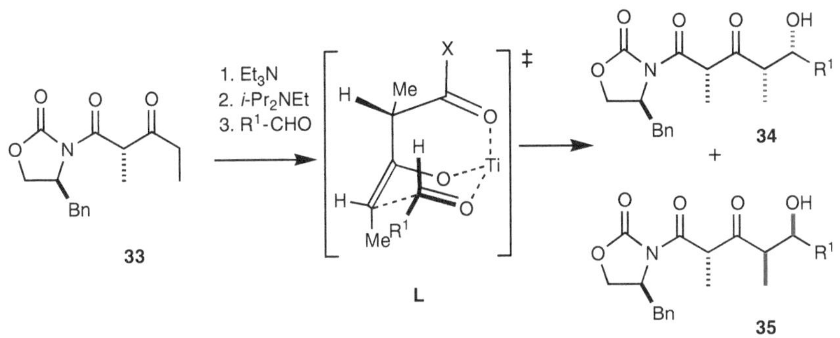

entry	R^1	yield (%)	**34 / 35**	**34 / 35**[a]
1	i-Pr	86	>99 / 1	5 / 95
2	$H_2C{=}CMe$	64	95 / 5	1 / 99
3	Et	86	>99 / 1	21 / 79
4	Ph	81	96 / 4	11 / 89

Scheme 2.4.14 Double stereodifferentiation in *syn*-selective aldol addition
[a]Reaction conditions: $Sn(OTf)_2$, Et_3N

An improvement over this protocol was published by Crimmins and coworker.[43]

For deployment of chiral benzyloxazolidine-2-thione in enantioselective aldol reaction see Franck et al.[44]

For the development and application of chiral oxadiazinones in enantioselective titanium enolate aldol reactions see Casper et al. and others.[45]

Chiral titanium(IV) alkoxides were used for the synthesis of optically active β-hydroxy ketones. Duthaler and coworkers generated chiral titanium enolates by transmetallation of lithium enolates of propionate with $CpTi(OR^2)_2Cl$ (R^2: 1, 2:5,6-di-O-isopropylidene-α-D-glucofuranose). Subsequent aldol additions with aldehydes afforded aldol adducts with a high degree of enantioselectivity (Scheme 2.4.15).[46]

These titanium enolate aldol methodologies have been applied in numerous syntheses of natural products.[47,48,49,50,51] Some spectacular examples are illustrated in the following schemes in order to demonstrate the power and selectivity of these transformations.

An example demonstrating the high preference for *syn*-diastereoselectivity can be found in the total synthesis of denticulatin B (Scheme 2.4.16).[52] For a boron enolate approach to denticulatin see De Brabander and Oppolzer[53] and via allylboranes see also Andersen et al.[54]

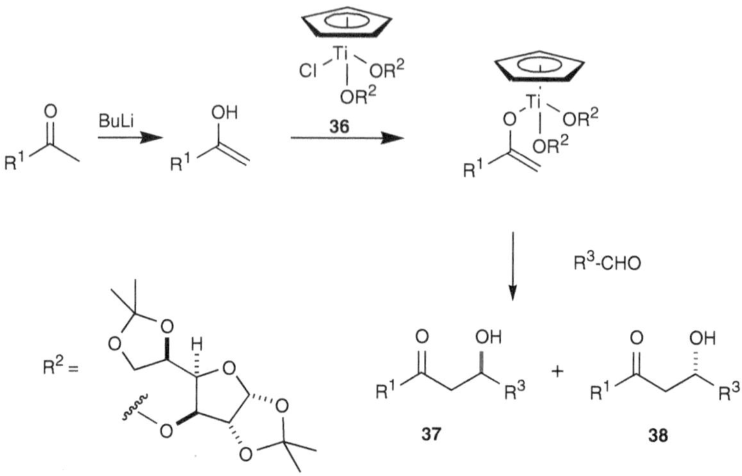

entry	R^1	R^3	yield (%)	**37** / **38**
1	*t*-BuO	*n*-Pr	51	97 / 3
2	*t*-BuO	*n*-C_7H_{15}	87	98 / 2
3	*t*-BuO	*i*-Pr	66	98 / 2
4	*t*-BuO	*t*-Bu	80	96 / 4
5	*t*-BuO	$H_2C = CMe$	81	98 / 2
6	*t*-BuO	Ph	69	98 / 2
7	*t*-BuO	2-furyl	62	95 / 5

Scheme 2.4.15 Stereoselective acetate aldol addition involving chiral titanium enolates

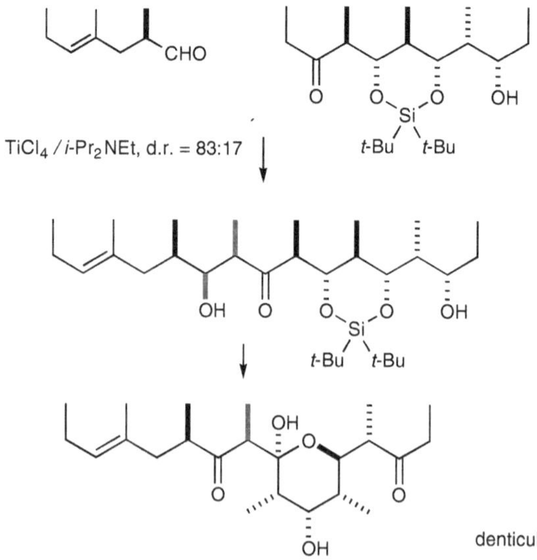

39

40

denticulatin B

Scheme 2.4.16 Total synthesis of denticulatin B

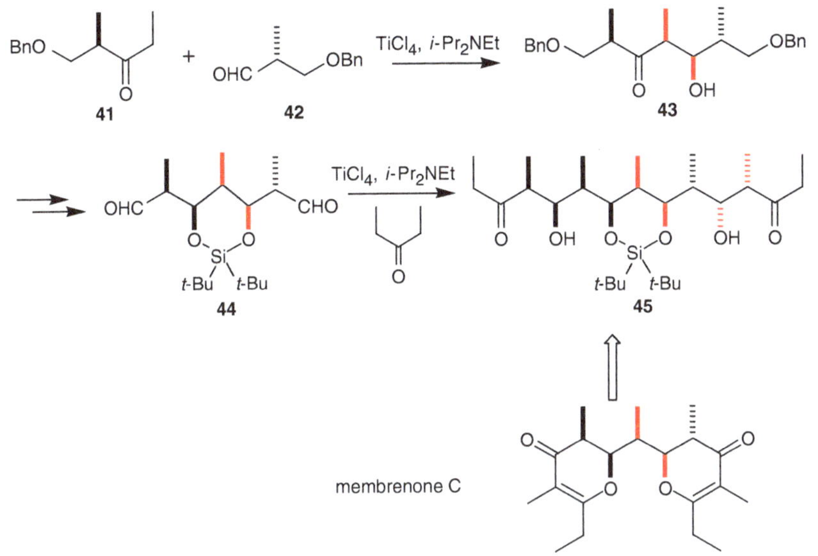

Scheme 2.4.17 Total synthesis of membrenone C

A similar reaction is the total synthesis of membrenone. A titanium enolate aldol step was used successfully two times during this sequence. The aldol reaction steps were performed in the presence of titanium(IV) chloride and diisopropylethylamine. The aldol products were observed with a high degree of *syn*-diastereoselectivity (aldol adduct **43**:d.r. = 98:2; diketone **45**:d.r. = 95:5) (Scheme 2.4.17).[55]

As a further example, the total synthesis of rapamycin by Danishefsky and coworkers demonstrates the utility of the application of titanium(IV) halogen alkoxides in aldol additions. At a very late stage in this synthesis, the cyclization of

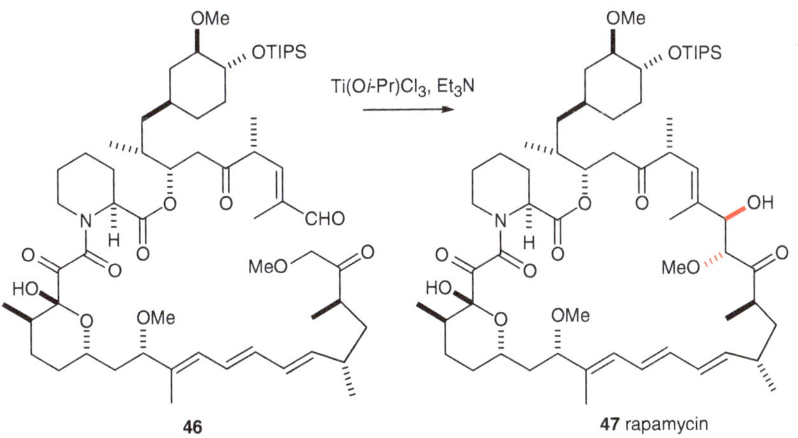

Scheme 2.4.18 Total synthesis of rapamycin

acyclic ketoaldehyde **46** was accomplished by an aldol addition in the presence of Ti(O*i*-Pr)Cl_3 and triethylamine. The cyclized *anti*-configured product **47** was isolated with 11% yield, together with 22% of the undesired *syn*-configured aldol product. This result once again underlines the described *syn*-selectivity generally observed in $TiCl_4$/amine-mediated aldol additions (Scheme 2.4.18).[56]

An application of chiral titanium enolates was reported in total synthesis of epothilone 490 by Danishefsky and coworkers (Scheme 2.4.19).[57] In their approach,

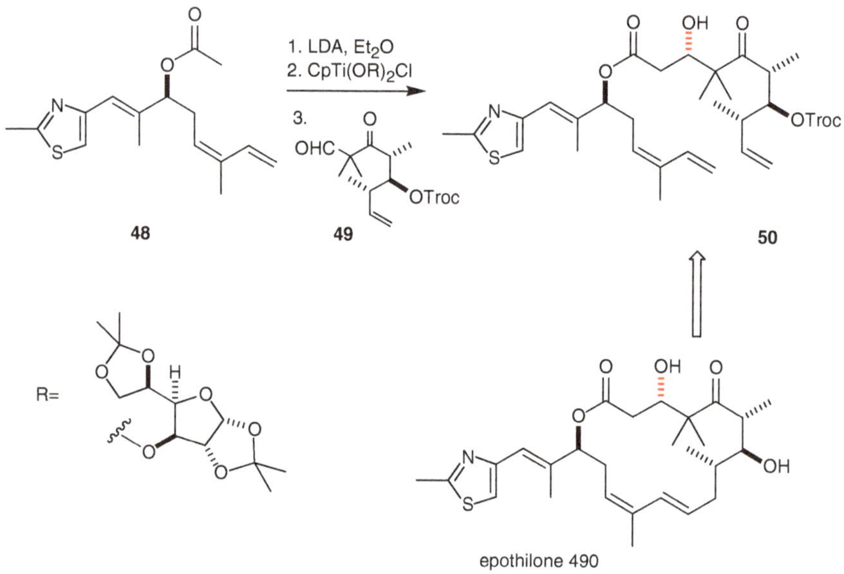

Scheme 2.4.19 Total synthesis of epothilone using chiral titanium enolates

they used the well-established Duthaler titanium enolate aldol methodology. For a comprehensive overview of this method see Duthaler and Hafner.[58] The generation of chiral titanium enolate was achieved by transmetallation of lithium enolate of acetate **48** with $CpTi(OR)_2Cl$ (R = 1, 2:5,6-di-O-isopropylidene-α-D-glucofuranose).

The aldol addition of chiral titanium enolate with chiral aldehyde **49** resulted aldol adduct **50** as a single isomer with 85% yield. Finally, ring closing metathesis yielded epothilone 490 with 64% yield.

One of the most frequently employed and most reliable aldol additions in natural product synthesis is the reaction of chiral titanium enolates, which are generated from amino acid-derived oxazolinone chiral auxiliaries. This method is based on the initial findings of Thornton and coworkers described above.[59] Crimmins used this method to synthesize callystatin A. In his approach, aldol additions with chiral titanium enolates were applied twice (Scheme 2.4.20).[60] The latter were generated by the use of $TiCl_4$ and (–)-sparteine, known as Crimmins'

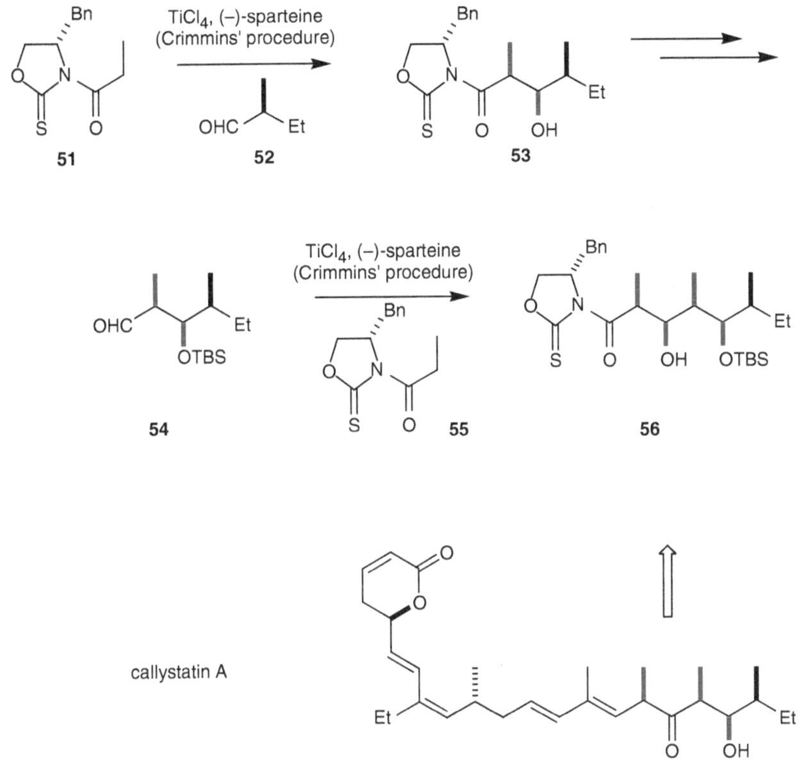

Scheme 2.4.20 Total synthesis of callystatin A

procedure. The reaction with (*S*)-2-methylbutanal **52** yielded the *syn*-aldol adduct **53** with 98% selectivity. Chain elongation was then performed by the same method, a second aldol addition of the chiral aldehyde **54** with the chiral titanium enolate of **55**. The final stereopentad **56** was isolated with 98% selectivity. For an aldol approach using tin enolates as well as titanium enolates leading to callystatin A and 20-*epi*-callystatin A see Enders et al.[61] For an aldol approach to callystatin A using the boron enolate chemistry see also Dias and Meira.[62] For an aldol approach to callystatin A using lithium enolate chemistry see also Kalesse et al.[63] For an overview of total syntheses of callystatin A see also Kalesse and Christmann.[64]

Four years later, the Crimmins group published the total synthesis of spongistatin, employing the same conditions as described above. By reacting aldehyde **57** to the titanium enolate of **58**, they obtained diol **59** with a high degree of *syn*-selectivity (dr: 96/4) (Scheme 2.4.21).[65,66,67]

Chakraborty and coworkers developed a total synthesis of crocacin using the Crimmins' procedure in initial aldol steps. By reacting cinnamaldehyde with the titanium enolate of **60**, they were able to isolate the allyl alcohol **61** as a single

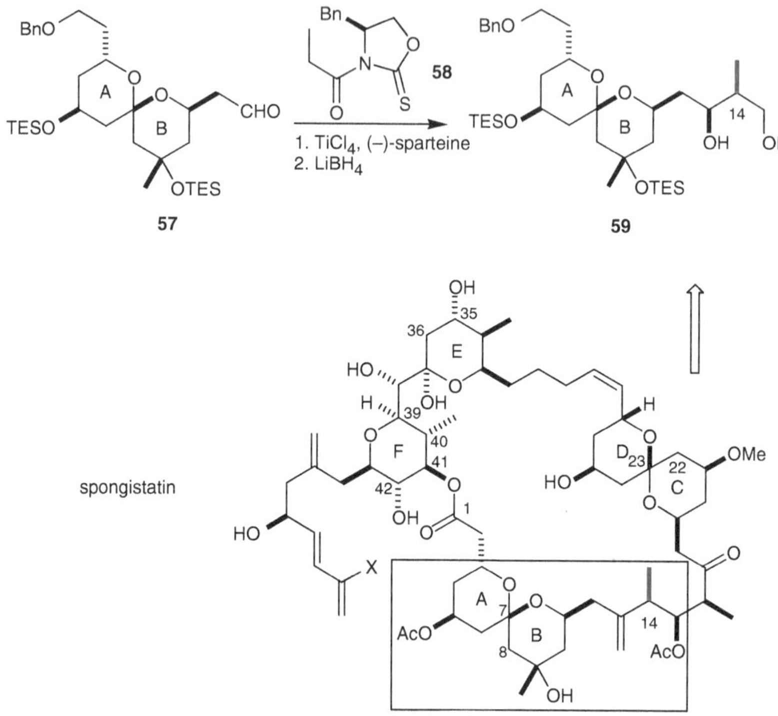

Scheme 2.4.21 Total synthesis of spongistatin

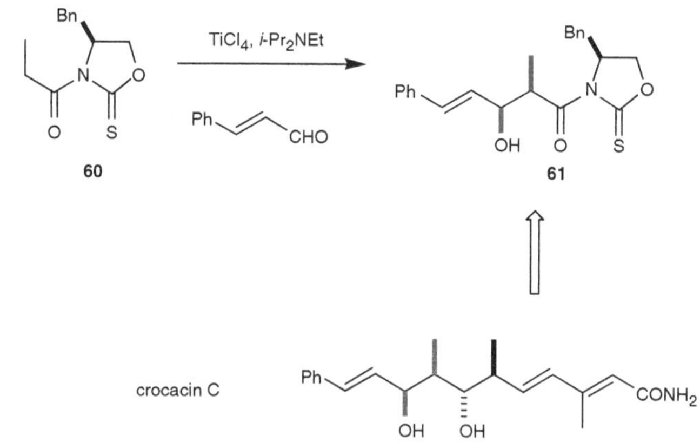

Scheme 2.4.22 Total synthesis of crocain C

syn-configured isomer (Scheme 2.4.22).[68] For an aldol approach to crocacin using boron enolate chemistry see also Dias and de Oliveira.[69] For the first total synthesis of crocacin using tin enolates see also Feutrill et al.[70]

As pointed out, the Gosh group developed several highly selective aldol methods based on aminoindanol chiral auxiliaries (see references[28–33]). In the first total

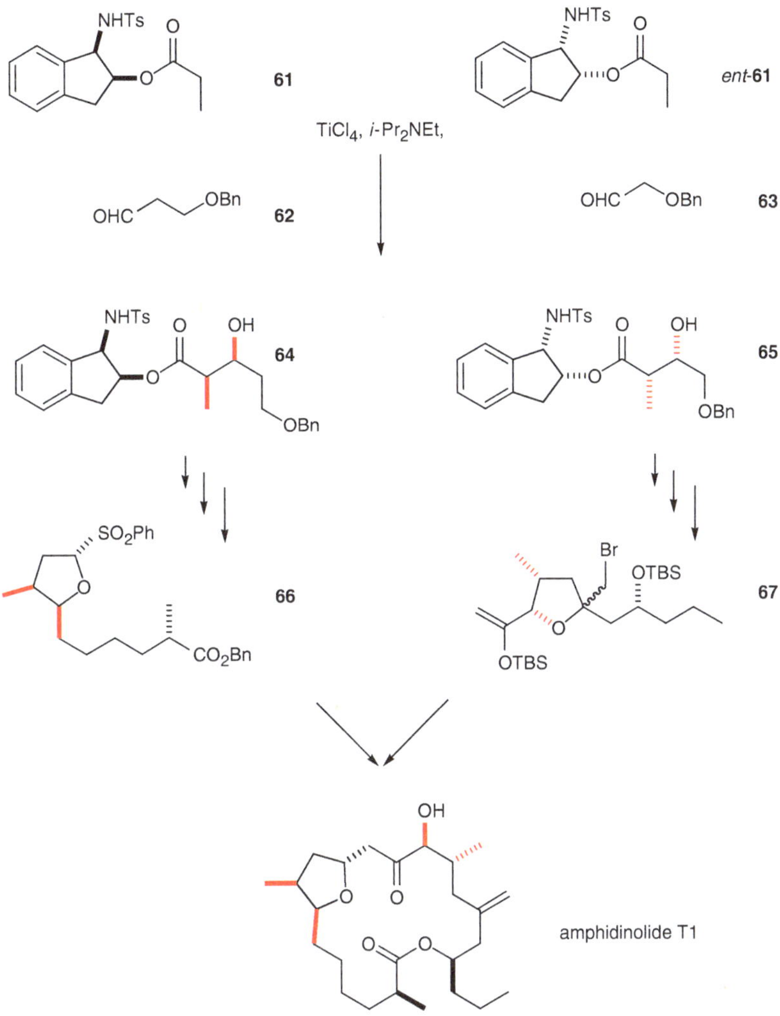

Scheme 2.4.23 Total synthesis of amphidinolide T1

synthesis of amphidinolide T1 they have demonstrated the usefulness of their methods (Scheme 2.4.23).[71] The construction of the two key intermediates **66** and **67** was performed according to the Gosh method. The aldol reaction of the titanium enolate of chiral ester **61** with 3-benzyloxypropionaldehyde **62** yielded a single *syn*-configured stereoisomer **64**. Alternatively, a single isomer of the *syn*-configured isomer **65** can be isolated by reacting the titanium enolate of ester *ent*-**61** with benzyloxyacetaldehyde **63**. Further transformations of the fragments **66** and **67** finally gave amphidinolide T1. These stereochemical results are explained by the transition states illustrated in Scheme 2.4.8. They have been rationalized by the use of a chelation-controlled model. *anti*-Diastereoselectivity is explained by a seven-membered transition state **E**. Whereas the excellent *syn*-diastereoselectivity that is observed in the reaction with α-oxygen-containing aldehydes can be explained by the transition state **D**.

For a comparative and comprehensive study of titanium enolates to boron enolates of *N*-acetylthiazolidinethiones in aldol additions during total synthesis of hennoxazole A see Smith et al.[72]

References

1. Reetz MT (2002) In: Organometallics in Synthesis – A Manual, Schlosser M (ed). Wiley, New York, 817
2. Mashima K (1997) In: Synthesis of Organometallic Compounds, Komiya, S. (ed). Wiley, New York
3. Reetz MT, Peter R (1981) Tetrahedron Lett 22:4691
4. Harrison C (1987) Tetrahedron Lett 28:4135
5. Evans DA, Rieger DL, Bilodeau MT, Urpi F (1991) J Am Chem Soc 113:1047
6. Rodriguez-Cisterna V, Villar C, Romea P, Urpi F (2007) J Org Chem 72:6631
7. Pellicena M, Solsona JG, Romea P, Urpi F (2008) Tetrahedron Lett 49:5265
8. Annunziata R, Cinquini M, Cozzi F, Cozzi PG, Coslandi E (1991) Tetrahdedron Lett 47:7897
9. Solsona JG, Nebot J, Romea P, Urpi F (2005) J Org Chem 70:6533
10. (a) Chakraborty TK, Goswami RK (2006) Tetrahedron Lett 47:4917; (b) Chakraborty TK, Goswami RK (2007) Tetrahedron Lett 48:6463
11. Murphy PJ, Procter G, Russell AT (1987) Tetrahedron Lett 28:2037
12. Annunziata R, Cinquini M, Cozzi F, Borgia AL (1992) J Org Chem 57:6339
13. Kazmaier U, Grandel R (1995) Synlett 945
14. Mahrwald R (1995) Tetrahedron 51:9015
15. Yamago S, Machii D, Nakamura E (1991) J Org Chem 56:2098
16. Evans DA, Trotter BW, Barrow JC (1997) Tetrahedron 53:8779
17. Nerz-Stormes M, Thornton ER (1986) Tetrahedron Lett 27:897
18. Shirodkar S, Nerz-Stormes M, Thornton ER (1990) Tetrahedron Lett 31:4699
19. Nerz-Stormes M, Thornton ER (1991) J Org Chem 56:2489
20. Evans DA, Rieger DL, Bilodeau MT, Urpi F (1991) J Am Chem Soc 113:1047
21. Bonner MP, Thornton ER (1991) J Am Chem Soc 113:1299
22. Ahn KH, Lee S, Lim A (1991) J Org Chem 57:5065
23. Yan T, Lee H, Tan C (1993) Tetrahedron Lett 34:3559
24. Yan T, Tan C, Lee H, Lo H, Huang T (1993) J Am Chem Soc 115:2613
25. Xiang Y, Olivier E, Quimet N (1992) Tetrahedron Lett 33:457
26. Ahn KH, Yoo DJ, Kim JS (1992) Tetrahedron Lett 33:6661
27. Casper DM, Burgeson JR, Esken JM, Ference GM, Hitchcock SR (2002) Org Lett 4:3739
28. Gosh AK, Fidanze S, Onishi M, Hussain KA (1997) Tetrahedron Lett 38:7171
29. Gosh AK, Kim J (2001) Tetrahedron Lett 42:1227
30. Gosh AK, Kim J (2002) Tetrahedron Lett 43:5621
31. Ghosh AK, Onishi M (1996) J Am Chem Soc 118:2527
32. Ghosh AK, Bischoff A (2004) Eur J Org Chem 2131
33. Ghosh AK, Li J, Perali RS (2006) Synthesis 3015
34. Crimmins MT, King BW, Tabet EA (1997) J Am Chem Soc 119:7883
35. Crimmins MT, Choudhary K (2000) Org Lett 2:775
36. Crimmins MT, King BW, Tabet EA, Choudhary K (2001) J Org Chem 66:894
37. Hajra S, Giri AK, Karmakar A, Khatu S (2007) Chem Commun 2408
38. Siegel C, Thornton ER (1986) Tetrahedron Lett 27:457

39. Siegel C, Thornton ER (1989) J Am Chem Soc 111:5722
40. Figueras S, Martin R, Romea P, Urpi F, Vilarassa J (1997) Tetrahedron Lett 38:1637
41. Evans, DA, Clark JS, Metternich R, Nowack VJ, Sheppard GS (1990) J Am Chem Soc 112:866
42. Evans DA, Dart M, Duffy JL, Rieger DL (1995) J Am Chem Soc 117:9073
43. (a) Crimmins MT, She J (2004) Synlett 1371; (b) Hodge MB, Olivo HF (2004) Tetrahedron 60:9397
44. Franck X, Seon-Meniel B, Figadere B (2006) Angew Chem Int Ed 45:5174
45. (a) Casper DM, Burgeson JR, Esken JM, Ferrence GM, Hitchcock SR (2002) Org Lett 4:3739; (b) Hoover T, Hitchcock SR (2003) Tetrahedron: Asymm 14:3233; (c) Hitchcock SR, Casper DM, Vaughn JF, Finefield JM, Ferrence GM, Esken JM (2004) J Org Chem 69:714; (c) Vaughn JF, Hitchcock SR (2004) Tetrahedron: Asymm 15:3449; (d) Squire MD, Davis RA, Chianakas KA, Ferrence GM, Standard JM, Hitchcock SR (2005) Tetrahedron: Asymm 16:1047; (e) Hoover TR, Groeper JA, Parrott RW, Chandrashekar SP, Finefield JM, Dominguez A, Hitchcock SR (2006) Tetrahedron: Asymm 17:1831; (f) Dore DD, Burgeson JR, Davis RA, Hitchcock SR (2006) Tetrahedron: Asymm 17:2386
46. Duthaler RO, Herold P, Wyler-Helfer S, Riedicker M (1990) Helv Chim Acta 28:497
47. Crimmins MT, Caussanel F (2006) J Am Chem Soc 128:3128
48. Gosh AK, Shevlin M (2004) In: Modern Aldol Reactions, Mahrwald R (ed). Wiley-VCH, Weinheim
49. Yeung KS, Paterson I (2005) Chem Rev 105:4237
50. Schetter B, Mahrwald R (2006) Angew Chem Int Ed 45:7506
51. Halim R, Brimble MA, Merten J (2006) Org Biomol Chem 4:1387
52. Paterson I, Perkins MV (1996) Tetrahedron 52:1811
53. De Brabander J, Oppolzer W (1997) Tetrahedron 53:9169
54. Andersen MW, Hildebrandt B, Hoffmann RW (1991) Angew Chem Int Ed Engl 30:97
55. Perkins MV, Sampson RA (2001) Org Lett 3:123
56. Hayward CM, Yohannes D, Danishefsky SJ (1993) J Am Chem Soc 115:9345
57. (a) Biswas K, Lin H, Njardarson JT, Chappell MD, Chou TC, Guan Y, Tong WP, He L, Horwitz SB, Danishefsky SJ (2002) J Am Chem Soc 124:9825; (b) Chappell MD, Stachel SJ, Lee CB, Danishefsky SJ (2000) Org Lett 2:1633; (c) Lee CB, Wu Z, Zhang F, Chappell MD, Stachel SJ, Chou TC, Guan Y, Danishefsky SJ (2001) J Am Chem Soc 123:5249; (d) Wu Z, Zhang F, Danishefsky SJ (2000) Angew Chem Int Ed Engl 39:4505
58. Duthaler RO, Hafner A (1992) Chem Rev 92:807
59. Nerz-Stormes M, Thornton ER (1986) Tetrahedron Lett 27:897
60. Crimmins MT, King BW (1998) J Am Chem Soc 120:9084
61. Enders D, Vicario JL, Job A, Wolberg M, Müller M (2002) Chem Eur J 8:4272
62. Dias LC, Meira PRR (2005) J Org Chem 70:4762
63. Kalesse M, Chary KP, Quitschalle M, Burzlaff A, Kasper C, Scheper T (2003) Chem Eur J 9:1129
64. Kalesse M, Christmann M (2002) Synthesis 981
65. Crimmins MT, Katz JD, Washburn DG, Allwein SP, McAtee LF (2002) J Am Chem Soc 124:5661
66. Crimmins MT, Katz JD, McAtee LC, Tabet EA, Kirincich SJ (2001) Org Lett 3:949
67. Crimmins MT, Katz JD (2000) Org Lett 2:957
68. Chakraborty TK, Jayaprakash S, Laxman P (2001) Tetrahedron 57:9461
69. Dias LC, de Oliveira LG (2001) Org Lett 3:3951
70. Feutrill JT, Lilly MJ, Rizzacasa MA (2000) Org Lett 2:3365
71. Gosh AK, Liu CJ (2003) J Am Chem Soc 125:2374
72. Smith TE, Kuo WH, Balskus EP, Bock VD, Roizen JL, Theberge AB, Carroll KA, Kurihara T, Wessler JD (2008) J Org Chem 73:142

2.5 Tin Enolates

Tin(II) enolates of ketones, esters or thioesters are accessible by reactions of tertiary amines and tin(II) triflate. High *syn*-diastereoselectivities were observed in reactions of aldehydes with tin(II) enolates of ketones. In pioneer experiments Mukaiyama and coworkers deployed chiral diamines **A**, **B** and **C** in enantioselective aldol additions.[1,2] The *syn*-configured aldol adducts were isolated with good enantioselectivities. Furthermore, when used with aromatic ketones enantioselectivities up to 90% were detected (Scheme 2.5.1).

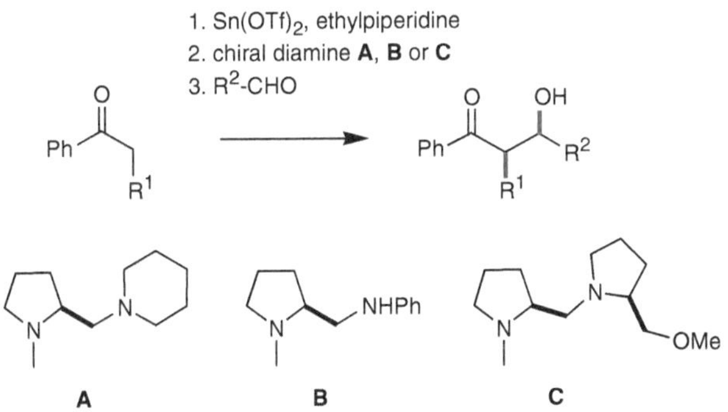

entry	R^1	R^2	chiral diamine	yield (%)	*syn* / *anti*	ee (%,*syn*)
1	Me	Ph	A	74	86 / 14	80
2	Me	4-MeC_6H_4	A	72	89 / 11	80
3	Me	4-ClC_6H_4	A	72	86 / 14	85
4	Me	4-MeOC_6H_4	A	78	89 / 11	80
5	Et	Ph	A	72	83 / 17	75
6	H	Ph	A	35	n.r.	75
7	Me	*i*-Pr	B	69	>95 / 5	75
8	Me	*t*-Bu	A	57	>95 / 5	90
9	Me	Cy	A	67	80 / 20	80

Scheme 2.5.1 Chiral diamines and tin(II) triflate in enantioselective Mukaiyama reaction

R. Mahrwald, *Aldol Reactions*,

When carboxylic acid derivatives, amides or thioamides were treated with aliphatic aldehydes under these conditions enantioselectivities up to 90% were obtained in the presence of chiral amine **A**. Representative examples are given in Scheme 2.5.2.[3,4] For a comprehensive overview of this development see also Mukaiyama and Kobayashi.[5]

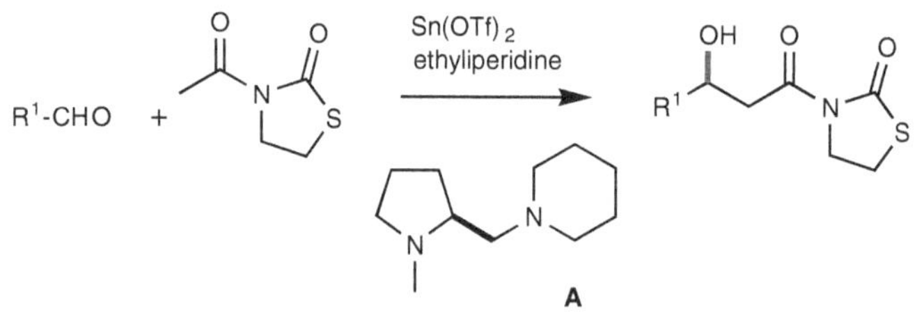

entry	R^1	yield (%)	ee (%)
1	Ph	79	65
2	*t*-Bu	63	>90
3	$PhCH_2CH_2$	76	>90
4	Cy	81	88
5	Et	70	90
6	*n*-$CH_3(CH_2)_4$	65	90

Scheme 2.5.2 Enantioselective aldol reaction of 3-acetylthiazolidine-2-thione

For an enantioselective aldol addition of tin enolates catalysed by chiral BINAP–silver triflate complexes see Yanagisawa et al.[6]

Applications of tin enolate in aldol addition were reported by Kirk and coworkers. They deployed tin(II) enolates of oxazolidinones in total synthesis of chiral fluorinated norepinephrines.[7]

An application of tin enolate aldol reactions in polyketide synthesis was described by Calter and coworkers.[8] An access to *anti-syn*-configured dipropionate equivalents was elaborated. These specific configured units were not attainable by well-established boron or titanium enolate aldol methodologies. By the use of tin enolates they were able to isolate the *anti-syn*-configured dipropionate units **1** (Scheme 2.5.3).

Evans and coworkers used an aldol addition of tin enolates in total synthesis of calyculin A. Tin(II) enolate of glycolate imide **3** was generated with the help of tin(II) triflate in the presence of triethylamine to construct the ketide structure of the north part of calyculin A. The observed unusual *anti*-selectivity was achieved by an additional treatment of the tin(II) enolate of the imide **3** with TMEDA (Scheme 2.5.4).[9] For an aldol approach to calyculin A using lithium enolate methodology see also Scarlato et al.[10]

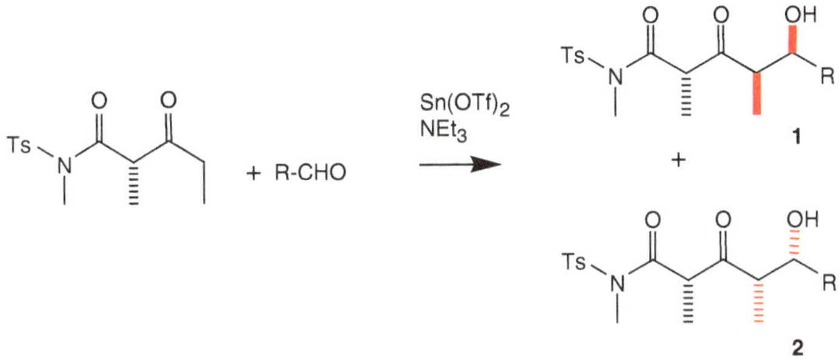

entry	R	yield (%)	ratio **1** / **2**
1	*i*-Pr	77	96 / 4
2	*n*-Pr	66	85 / 15
3	Ph	65	86 / 14
4	$BnOCH_2$–	47	80 / 20
5	$TBSO(CH_2)_2$–	61	89 / 11

Scheme 2.5.3 Aldol additions of chiral tin(II) enolates

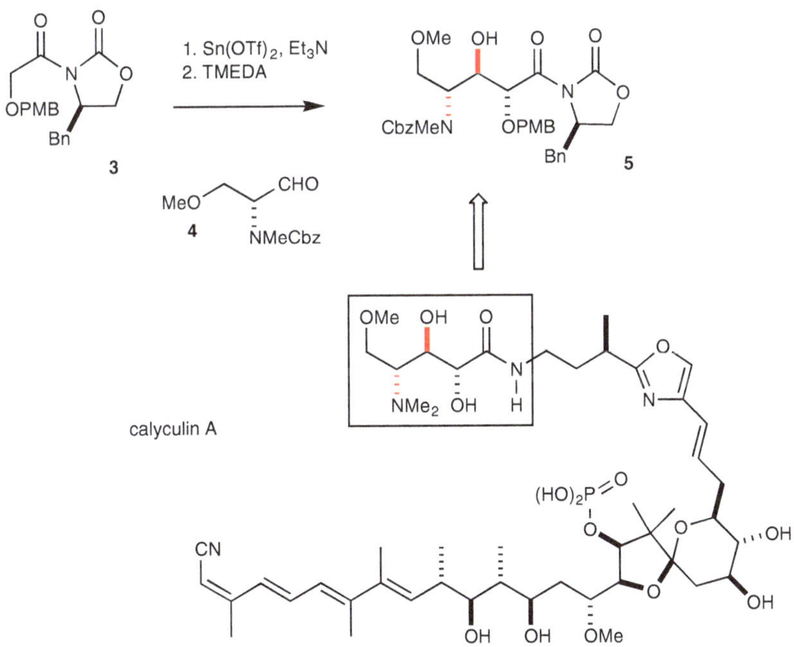

Scheme 2.5.4 Evans' total synthesis of calyculin A

Very recently, the same procedure was used again by Evans and coworkers in the total synthesis of aflastatin A. Tin(II) triflate and triethylamine served to establish the desired *anti*-configuration in the starting chiral compound **8**. The subsequent boron enolate aldol addition resulted in the formation of exclusively the *anti-syn-anti* stereoarray found in the C33–C36 region of aflastatin A (Scheme 2.5.5).[11]

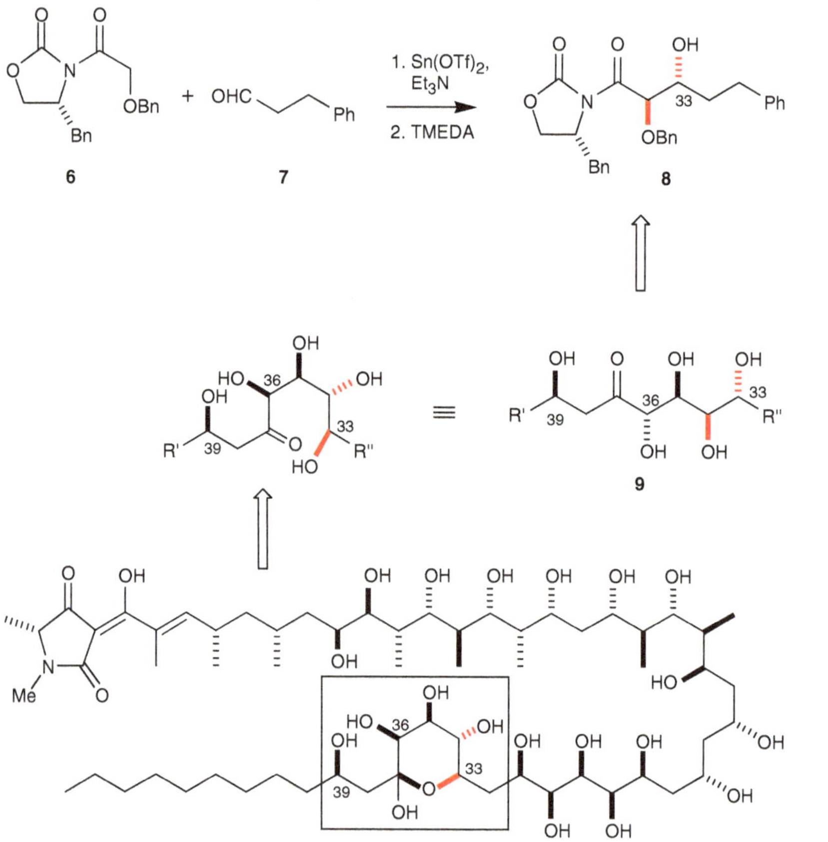

Scheme 2.5.5 Total synthesis of aflastatin A

During the total synthesis of oasomycin A the required configuration of the important C29–C37 subunit at C32 and C33 was installed by a tin enolate aldol step of the chiral ketoimide **11** to aldehyde **12**. A diastereomeric ratio of 95:5 was detected (Scheme 2.5.6).[12] The initial enantioselective aldol addition of diene **14** and benzyloxyacetaldehyde **15** was realized by a tridentate bis(oxazolinyl)pyridine (pybox)–Cu(II) complex to install the required configuration at C37.

During a convergent total synthesis of apoptolidine a highly stereoselective aldol addition was needed to combine ethylketone **17** with aldehyde **18** to yield

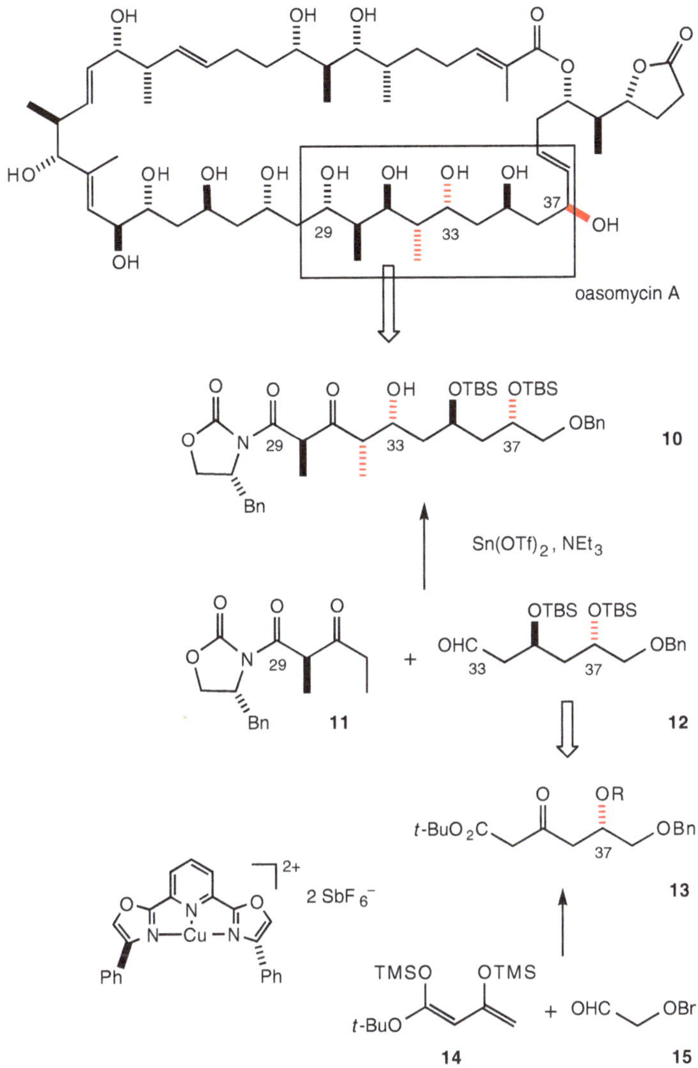

Scheme 2.5.6 Total synthesis of oasomycin A

required *syn*-configured stereotetrad **16** (Scheme 2.5.7).[13] The authors tested several *syn*-selective methods and among them tin(II) enolate aldol addition proved to be the procedure of choice. High diastereoselectivities were detected (80/8/5/0).

In the total synthesis of decarestrictine C2 a key step was realized by a tin(II) enolate aldol addition (Scheme 2.5.8).[14] The (*R*)-configured aldol adduct **19** of aldehyde **20** and chiral acetylthiazolidine-2-thione **21** was obtained with a high

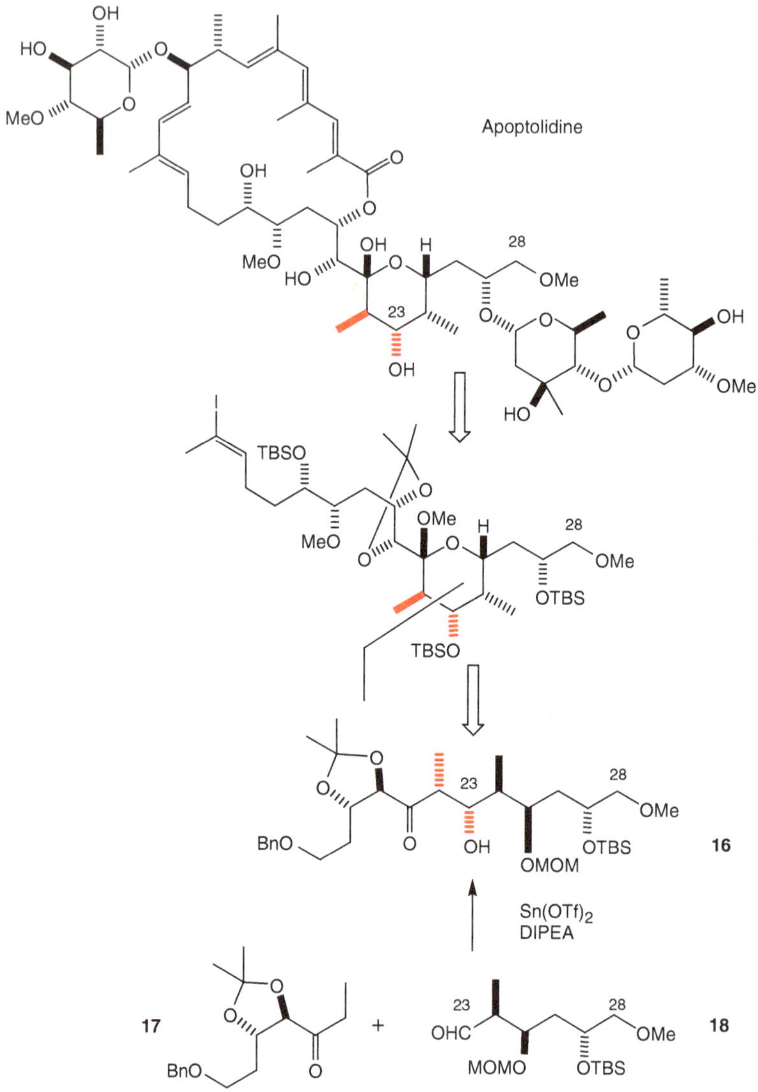

17 + 18 → 16

Scheme 2.5.7 Stereoselective tin enolate aldol step in total synthesis of apoptolidine

degree of diastereoselectivity (dr: 97/3). This result concurs with the fundamental findings of Nagao and coworkers. The authors described the first application of the chiral ketone **21** in tin enolate aldol additions.[15]

Tin enolate of chiral ethylketone **23** was reacted with cinnamaldehyde **22** in a total synthesis of crocacin C. High *syn*-diastereoselectivity was observed (dr: 99/1) (Scheme 2.5.9).[16]

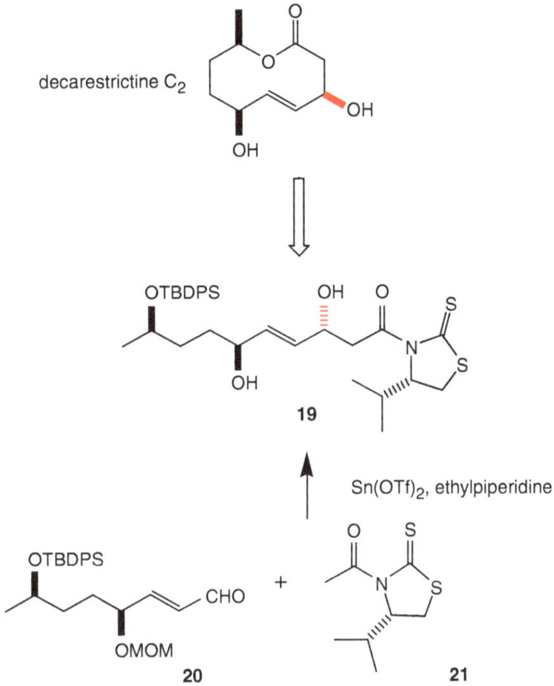

Scheme 2.5.8 Total synthesis of decarestrictine C_2

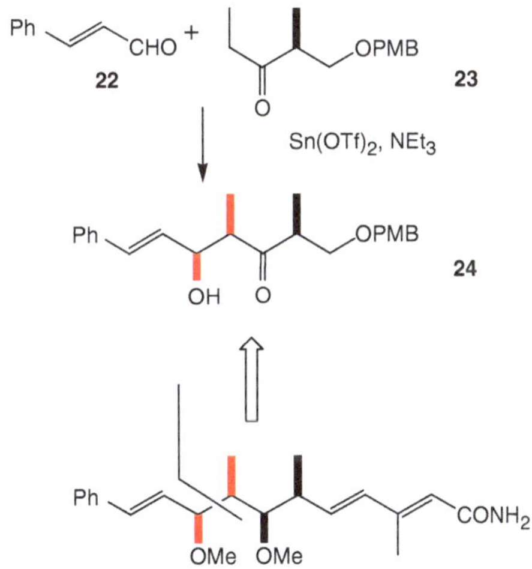

Scheme 2.5.9 Total synthesis of crocacin C

Woerpel and Peng reported a total synthesis of dihydroerythronolide A. By combining the aldehyde **25** with the tin enolate of ethylketone **26** they were able to isolate the important intermediate **27** with 90% yield and the required *syn*-configuration at C8–C9 with a ratio of 99:1 (Scheme 2.5.10).[17] By deploying of corresponding lithium enolates only poor selectivities were detected.

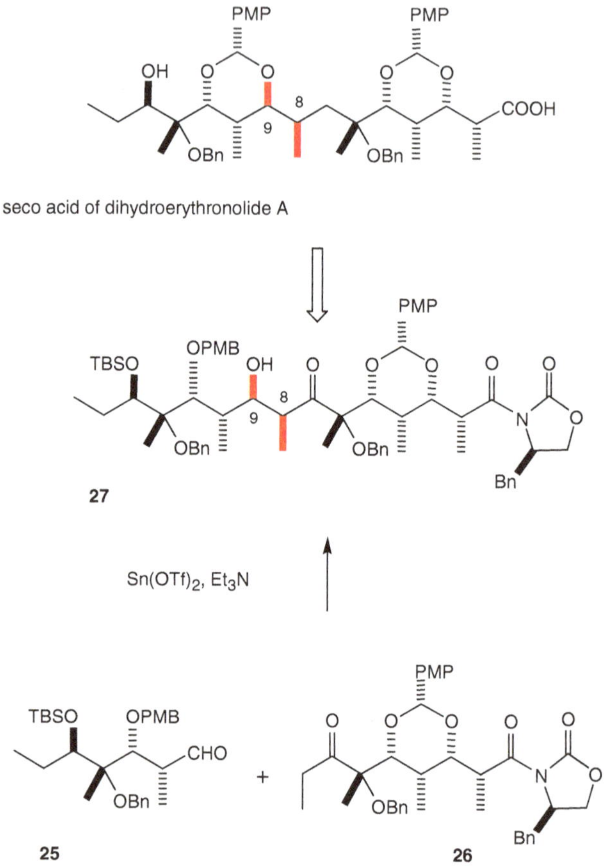

Scheme 2.5.10 Total synthesis of dihydroerythronolide A

For application of tin enolate aldol methodology in total synthesis of hennoxazole A see Smith et al.[18]

References

1. Mukaiyama T, Iwasawa N, Stevens, RW, Haga, T (1984) Tetrahedron 40:1381
2. Iwasawa N, Yura T, Mukaiyama T (1982) Chem Lett 1441
3. Iwasawa N, Yura T, Mukaiyama T (1989) Tetrahedron 45:1197
4. Iwasawa N, Mukaiyama T (1983) Chem Lett 297
5. Mukaiyama T, Kobayashi S (1994) Org React 46:1
6. Yanagisawa A, Matsumoto Y, Nakashima H, Asakawa K, Yamamoto H (1997) J Am Chem Soc 119:9319
7. Herbert B, Kim IH, Kirk KL (2001) J Org Chem 66:4892
8. Calter MA, Song W, Zhou J (2004) J Org Chem 69:1270

9. Evans DA, Gage JR, Leighton JL (1992) J Am Chem Soc 114:9434
10. Scarlato GR, DeMattei JA, Chong LS, Ogawa AK, Lin MR, Armstrong RW (1996) J Org Chem 61:6139
11. (a) Evans DA, Glorius F, Burch JD (2005) Org Lett 7:3331; (b) Evans DA, Trenkle WC, Zhang J, Burch JD (2005) Org Lett 7:3335
12. Evans DA, Nagorny P, Reynolds DJ, McRae KJ (2007) Angew Chem Int Ed 46:541
13. Abe K, Kato K, Arai T, Rahim MA, Sultana I, Matsumura S, Toshima K (2004) Tetrahedron Lett 45:8849
14. Arai M, Morita N, Aoyagi S, Kibayashi C (2000) Tetrahedron Lett 41:1199
15. (a) Nagao Y, Hagiwara Y, Kumagai T, Ochiai M, Inoue T, Hashimoto K, Fujita E (1986) J Org Chem 51:2391; (b) Nagao Y, Kumagai T, Nagase Y, Tamai S, Inoue Y, Shiro M (1992) J Org Chem 57:4232
16. Feutrill JT, Lilly MJ, White JM, Rizzacasa MA (2008) Tetrahedron 64:4880
17. Peng ZH, Woerpel KA (2003) J Am Chem Soc 125:6018
18. Smith TE, Kuo WH, Balskus EP, Bock VD, Roizen JL, Theberge AB, Carroll KA, Kurihara T, Wessler JD (2008) J Org Chem 73:142

3 Catalytic Aldol Additions

It was 1973 when Mukaiyama and coworkers described the first application of silyl enol ethers in aldol additions.[1] This reaction is promoted by Lewis acids and allowed a catalytic and an enantioselective execution for the first time.[2] Moreover, the regioselectivity can be controlled efficiently by using defined silyl enol ether of unsymmetrical ketones. High chemoselectivities were observed by using aldehydes, ketones and carboxylic esters in these transformations (Scheme 3.1).[3,4]

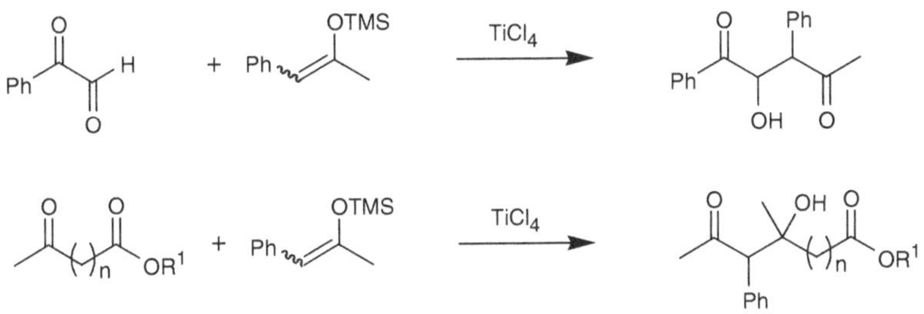

Scheme 3.1 Titanium(IV) chloride-mediated Mukaiyama reaction

The level and the sense of stereoselectivity often vary and depend on the aldehydes, silyl enol ethers and on the Lewis acids used. The stereochemical results have been rationalized by considering so-called open transitions states (Scheme 3.2).[5] This illustration is a simple working model based on repulsive forces only. The influence of different Lewis acids, the influence of different substituents and the fate of silyl group during the reaction were not involved in this model. For further discussion and mechanistic understanding see Hiraiwa et al. and others.[6]

These initial studies were the starting point of ongoing developments of new methods of the so-called Mukaiyama reaction. The full potential of this reaction by the use of different and chiral Lewis acids was realized by discoveries of several groups in the following time. A vast number of publications reported the development of new Lewis acids.[7] For several comprehensive overviews see Gennari and others.[8–16] For latest application of polymer-supported Lewis acids in Mukaiyama reaction see Doherty et al. and others.[17] For an overview of Mukaiyama reactions in aqueous media see Mlynarski and Paradowska.[18]

R. Mahrwald, *Aldol Reactions*,

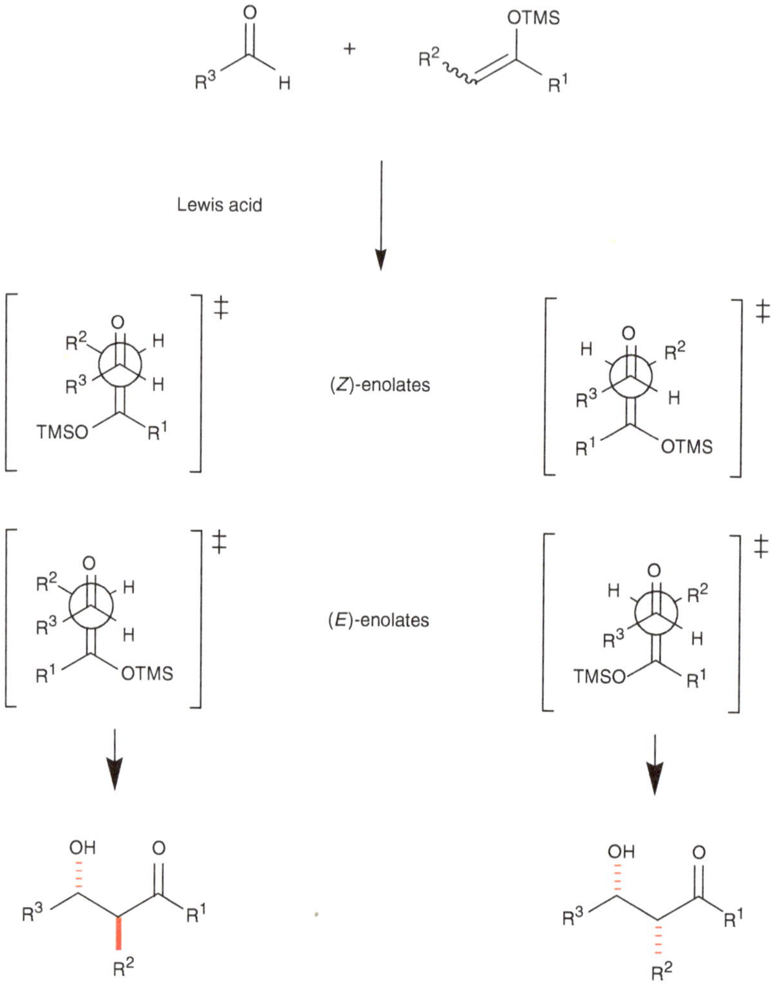

Scheme 3.2 Extended transition state models

References

1. Mukaiyama T, Narasaka K, Banno K (1973) Chem Lett 1011
2. Mukaiyama T, Banno K, Narasaka K (1974) J Am Chem Soc 96:7503
3. Banno K, Mukaiyama T (1975) Chem Lett 741
4. Banno K, Mukaiyama T (1976) Bull Chem Soc Jpn 49:2284
5. Murata S, Suzuki M, Noyori R (1980) J Am Chem Soc 102:3248
6. (a) Hiraiwa Y, Ishihara K, Yamamoto H (2006) Eur J Org Chem 1837; (b) Denmark SE, Lee W (2008) Chem Asian J 3:327; (c) Patel SG, Wiskur SL, (2009) Tetrahedron Lett 50:1164
7. (a) Yamamoto H, Ishihara K (eds) (2008) Acid Catalysis in Modern Organic Synthesis, Wiley, Weinheim; (b) Yamamoto H (ed) (2000) Lewis Acids in Organic Synthesis, Wiley, Weinheim

8. Gennari C (1993) In: Comprehensive Organic Synthesis, Trost BM, Fleming I, Heathcock CH (eds). Pergamon, Oxford, vol 2, p 629
9. Braun W (1996) In: Houben-Weyl, Methoden der Organischen Chemie, Helmchen G, Hoffmann RW, Mulzer J, Schaumann E (eds). Thieme, Stuttgart, vol E21b, p 1603
10. Nelson SG (1998) Tetrahedron: Asymm 9:357
11. Mukaiyama T, Matsuo JI (2004) In: Modern Aldol Reactions, Mahrwald R (ed). Wiley-VCH, Weinheim, vol. 1, p 137
12. Carreira EM, Fettes A, Marti C (2006) Org Reactions 67:1
13. Kalesse M (2005) Top Curr Chem 244:43
14. Mukaiyama T (1999) Tetrahedron 55:8609
15. (a) Gröger H, Vogl EM, Shibasaki M (1998) Chem Eur J 4:1137; (b) Johnson JS, Evans DA (2000) Acc Chem Res 33:325; (c) Machajewski TD, Wong CH (2000) Angew Chem Int Ed 39:1352; (d) Carreira EM (1999) In: Comprehensive Asymmetric Synthesis, Jacobsen EN, Pfaltz A, Yamamoto H (eds). Springer, Heidelberg, vol 3, p 998; (e) Carreira EM (2000) In: Comprehensive Asymmetric Synthesis, 2nd edn, Ojima I (ed). Wiley-VCH, New York, p 513.
16. Alcaide B, Almendros P (2002) Eur J Org Chem 1595
17. (a) Doherty S, Goodrich P, Hardacre C, Parvulescu V, Paun C (2008) Adv Synth Cat 350:295; (b) Fraile JM, Perez I, Mayoral JA (2007) J Cat 252:303; (c) Costantino U, Fringuelli F, Nocchetti M, Piermatti O (2007) Appl Cat A 326:100; (d) Gibson VC, Redshaw C, Solan GA (2007) Chem Rev 107:1745; (e) Gu Y, Ogawa C, Kobayashi J, Mori Y, Kobayashi S (2006) Angew Chem Int Ed 45:7217; (f) Fraile JM, Perez I, Mayoral JA, Reiser O (2006) Adv Synth Cat 348:1680
18. Mlynarski J, Paradowska J (2008) Chem Soc. Rev 37:1502

3.1 Lewis Acid-Catalysed Aldol Reactions

3.1.1 Titanium Lewis Acids

Mukaiyama and coworkers were the first to demonstrate the utility of titanium Lewis acids in aldol additions. In their initial experiments they used titanium(IV) chloride in catalytic amounts in aldol reactions with silyl enol ethers.[1,2,3] These initial findings were subsequently generalized and optimized in the following time. Numerous publications described the design of highly effective chiral titanium Lewis acids in enantioselective Mukaiyama reactions. Ab initio calculations were employed to investigate the mechanism of Lewis acid-catalysed Mukaiyama reactions.[4] This important development was the object of several comprehensive revi ews.[5,6,7,8,9,10,11,12,13,14,15,16] For that reasons only selected and latest examples should illustrate this development.

Optical active binaphthol is one of the most frequently applied ligands in the synthesis of chiral titanium Lewis acids. Reetz and coworkers were the first to report enantioselective Mukaiyama aldol reactions catalysed by modified BINOL–titanium(IV) complexes.[17] These findings were optimized for acetate aldol additions. When used with 5 mol% of this so-called Mikami catalyst[18] the expected β-hydroxy carboxylic esters were obtained with a high degree of enantioselectivity (Scheme 3.1.1.1).[19]

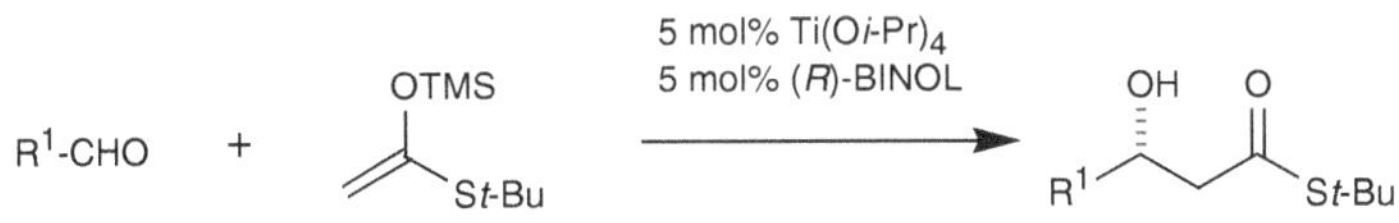

1

entry	R^1	yield (%)	ee (%)
1	Ph	71	96
2	EtO_2C	65	63
3	n-C_6H_{13}	76	93
4	$TBSO(CH_2)_5$	22	95
5	$(CH_2)_4CO_2n$-Pr	70	> 97

Scheme 3.1.1.1 BINOL–titanium-catalysed enantioselective acetate aldol addition

Later on Reissig and coworkers demonstrated the utility of 6.6′-dibrominated BINOL–titaniumisopropoxide in Mukaiyama reactions.[20] Soriente and coworkers applied chiral BINOL–titanium complexes to vinylogous Mukaiyama reactions.[21,22,23,24,25] Results of transformations with cyclic diene components are given in Scheme 3.1.1.2.

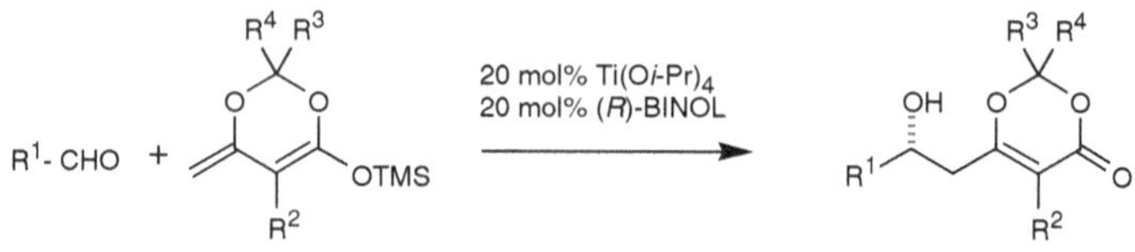

entry	R^1	R^2	R^3	R^4	yield (%)	ee (%)
1	n-C_9H_{19}	H	Me	Me	79	89 (*S*)
2	n-C_7H_{15}	H	Me	Me	90	92 (*S*)
3	$PhCH_2CH_2$	H	Me	Me	85	89 (*S*)
4	Ph	H	Me	Me	84	>99 (*R*)
5	3-furyl	H	Me	Me	75	>99 (*R*)
6	PhCH = CH	H	Me	Me	92	99 (*R*)
7	4-$MeOC_6H_4$	H	Me	Me	66	97 (*R*)
8	Ph	H	– $(CH_2)_5$ –		69	92 (n.r.)
9	3-furyl	Me	Me	Me	>90	90 (n.r.)

Scheme 3.1.1.2 BINOL–titanium-catalysed enantioselective vinylogous aldol addition

Also, acyclic diene enolate **2** was reacted in the presence of substoichiometric amounts of BINOL–titanium complexes with aldehydes to yield δ-hydroxy esters with a high degree of enantioselectivity (Scheme 3.1.1.3).[26,27] For studies of BINOL–titanium(IV) isopropoxide-complex in the presence of $B(OMe)_3$ see Heumann and Keck.[28]

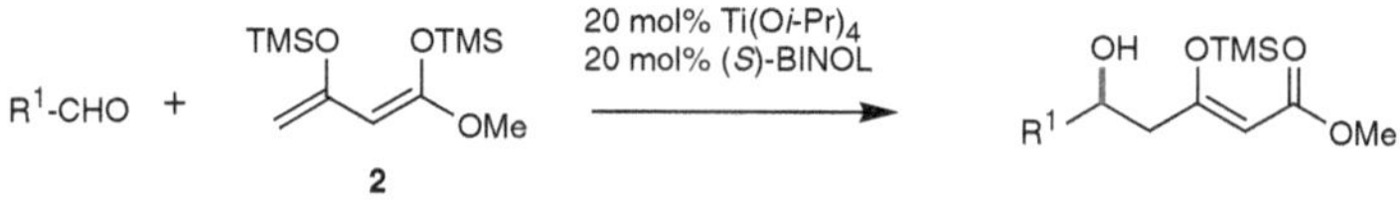

entry	R^1	yield (%)	ee (%)
1	n-C_9H_{19}	50	87
2	4-NO_2-C_6H_4	86	90
3	4-MeO-C_6H_4	54	93
4	Ph	33	95
5	3-furyl	48	83

Scheme 3.1.1.3 BINOL–titanium-catalysed enantioselective vinylogous aldol addition

Mikami and coworkers intensively studied the use of complexes derived of $TiCl_2(Oi\text{-}Pr)_2$ and BINOL. They were able to demonstrate the optional approach to optical active *syn*- or *anti*-configured aldol adducts **5** and **8** of glyoxylate **3** when used with *E*- or *Z*-enolsilanes of thioesters **4** and **7** (Scheme 3.1.1.4). [29,30,31]

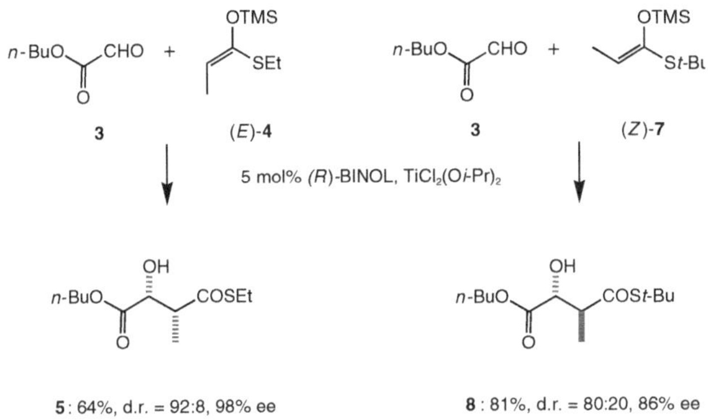

Scheme 3.1.1.4 Configurative results of (*Z*)- or (*E*)-enolates in Mukaiyama aldol additions

Also, Mikami and coworkers described the application of BINOL–titanium(IV) dichloride **9** in Mukaiyama reactions.[31] By using 5 mol% of the catalyst a silatropic ene reaction is observed. As a result of this mechanism the corresponding β-silyloxy ketones were obtained with a high degree of enantioselectivity. Results are given in Scheme 3.1.1.5.

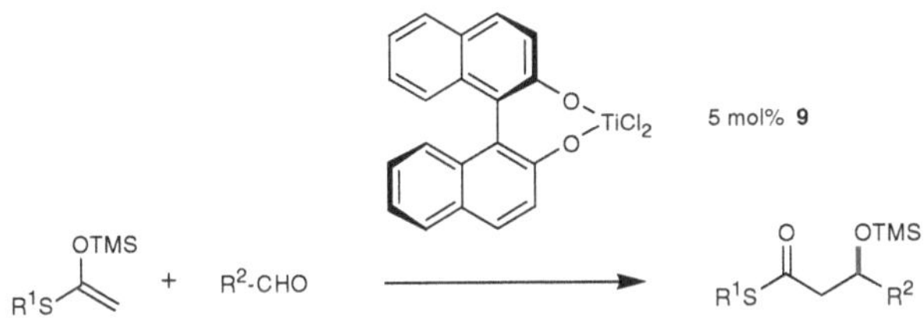

entry	R^1	R^2	yield (%)	ee (%)	configuration
1	Et	$BnOCH_2$	81	94	(*R*)
2	Et	$BnOCH_2$	96	85	(*R*)
3	*t*-Bu	$BnOCH_2$	80	96	(*S*)
4	*t*-Bu	CH_2Cl	61	91	(*R*)
5	Et	CH_2Cl	47	80	(*R*)
6	Et	CH_2NHBoc	64	88	(*R*)
7	*t*-Bu	C_8H_{17}	60	91	(*S*)
8	Et	C_8H_{17}	67	86	(*S*)
9	Et	$CHMe_2$	61	85	(*R*)
10	Et	*trans*-MeCH = CH	60	81	(*R*)

Scheme 3.1.1.5 BINOL titanium(IV) dichloride-catalysed Mukaiyama reaction

As illustrated in Scheme 3.1.1.5 this aldol process is going through a silatropic ene transformation. For investigations and discussion of this phenomenon during the Mukaiyama reaction see also references.[31,32,33]

Sato and coworkers applied the same chiral titanium complex **9** in vinylogous Mukaiyama reactions.[34] Results of these investigations are shown in Scheme 3.1.1.6. For a comprehensive overview of vinylogous Mukaiyama reactions see reference.[35]

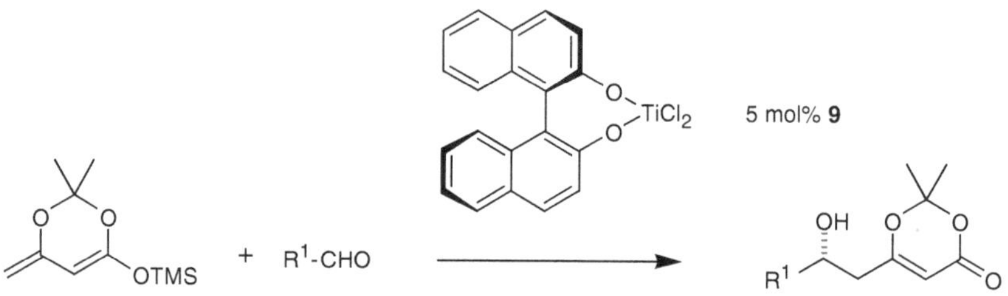

10

entry	R^1	yield (%)	ee (%)
1	*n*-Bu	38	92
2	Ph	55	88

Scheme 3.1.1.6 Enantioselective vinylogous Mukaiyama reaction

Carreira and coworkers developed a novel class of tridentate ligands for the synthesis of chiral titanium(IV) complexes. This catalyst represents chiral BINOL(IV) titanium complexes coordinating a Schiff base and they were used in acetate aldol additions.[36] They are unique and can be used with a variety of aldehydes – even with enolizable aldehydes. Dramatically enhanced yields, enantioselectivities and catalyst efficiency in Mukaiyama aldol additions were observed by optimization of the first generation of these complexes. Chiral tridentate chelate **11** was isolated by incorporation of 3.5-di-*tert*-butylsalicylic acid as a ligand. The aldol adducts were obtained with a high degree of enantioselectivity (Scheme 3.1.1.7). This catalytic system has found application in a series of natural product syntheses.[37]

The same authors described the application of the Schiff base titanium(IV) complex in vinylogous aldol additions.[38] Results of these investigations are shown in Scheme 3.1.1.8. The optically active β-hydroxy dioxinones represent useful masked aldol adducts.

Other chiral titanium Lewis acids and their application in Mukaiyama reactions incorporating chiral salen ligands as well as BINOL ligands were reported (chiral 1.2-diols,[39,40] calixarene[41,42] and mandelic acid[43]). An application of chiral BINOL-derived titanium complexes in direct and catalytic aldol additions is described in Schetter and Mahrwald.[44]

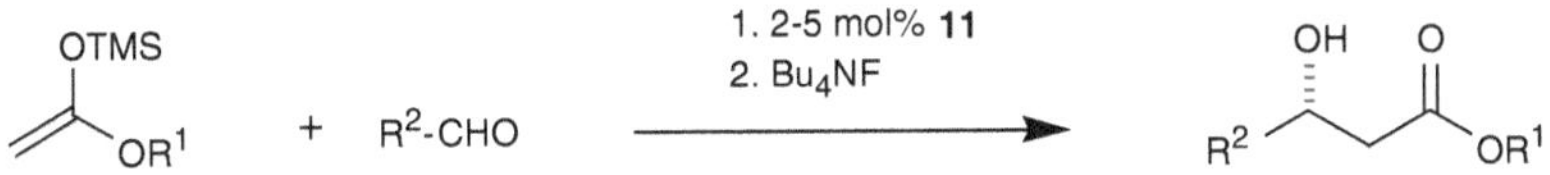

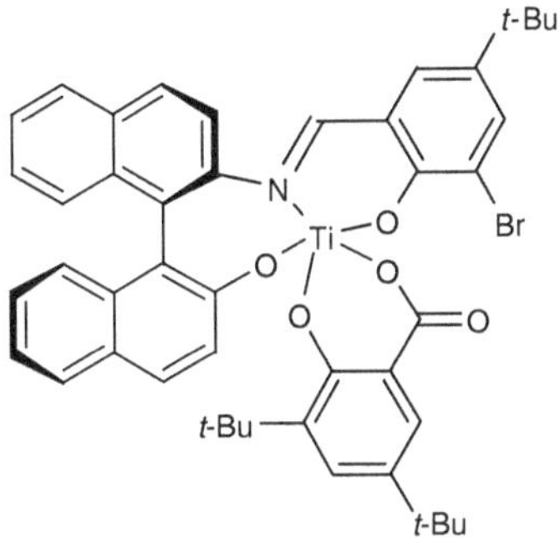

11

entry	R^2	ee (%)[a]	ee (%)[b]
1	*trans*-$CH_3CH = CH$	92	97
2	*n*-Pr	88	95
3	*trans*-PhHC = CH	93	97
4	$Ph(CH_2)_2$	89	94
5	Cy	94	95
6	Ph	93	96

Scheme 3.1.1.7 Carreira's catalyst in acetate Mukaiyama reactions
[a] R^1 = Et, 5 mol % catalyst; [b] R^1 = Me, 2 mol % catalyst

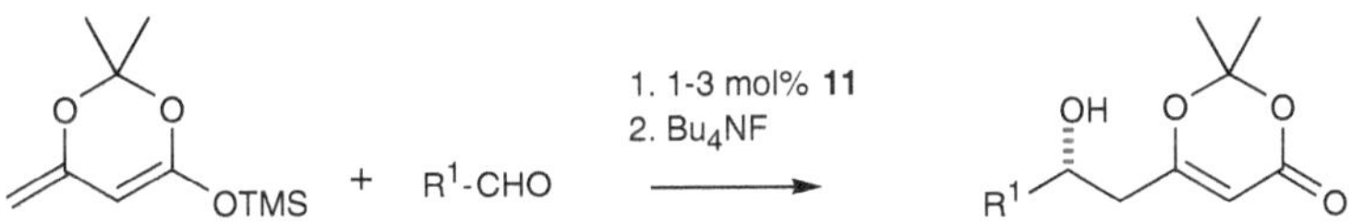

10

entry	R^1	yield (%)	ee (%)
1	*i*-$Pr_3SiC{\equiv}C$	86	91
2	*cis*-$TBSOCH_2HC = CH$	97	94
3	*trans*-PhHC = CH	88	92
4	Ph	83	84
5	$Ph(CH_2)_2$	97	80
6	*trans*-$Bu_3SnHC = CH$	79	92

Scheme 3.1.1.8 Carreira's catalyst in vinylogous aldol additions

Interesting investigations were conducted by Feng and coworkers.[45] They used a series of tridentate titanium(IV) Schiff base complexes in aldol reactions with vinylogous silyl enol ethers. The authors were able to demonstrate that the reactions at higher temperature are undergoing through a highly enantioselective Diels–Alder process (94% ee), whereas at −78°C a less enantioselective Mukaiyama aldol process was observed (30% ee).

Rychnovsky et al. used titanium(IV) complex **11** in a total synthesis of the polyene macrolide roflamycoin (Scheme 3.1.1.9). The absolute configuration of C29 in the C35–C29 segment **14** of roflamycoin was established by the use of a chiral titanium(IV) complex **11** in an acetate Mukaiyama aldol reaction.[46]

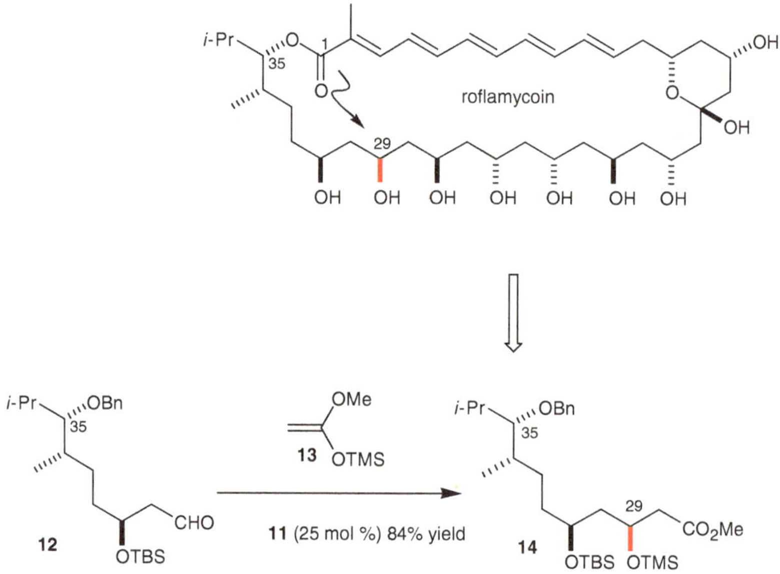

Scheme 3.1.1.9 Total synthesis of roflamycoin

A further application of the Carreira catalyst **11** can be found in the total synthesis of macrolactin A. By using the (*S*)-configured titanium catalyst (*S*)-**11**, the C3–C7 unit (*S*)-**16** was obtained in 92% ee and also, by applying (*R*)-configured titanium catalyst (*R*)-**11** the C11–C15 segment (*R*)-**16** was constructed. Both segments (*S*)-**16** and (*R*)-**16** were obtained from the same aldehydes **15** and diene **10** with same yields and enantioselectivities (Scheme 3.1.1.10).[47]

In a total synthesis of dermostatin Keck's BINOL–titanium(IV) complex was applied in aldol additions to establish configuration at C31 in aldol adduct **18**, albeit with modest yields and poor diastereoselectivity. By a comparative application of tin enolate aldol addition high yields and diastereoselectivities of the corresponding aldol adduct **19** were observed (Scheme 3.1.1.11).[48]

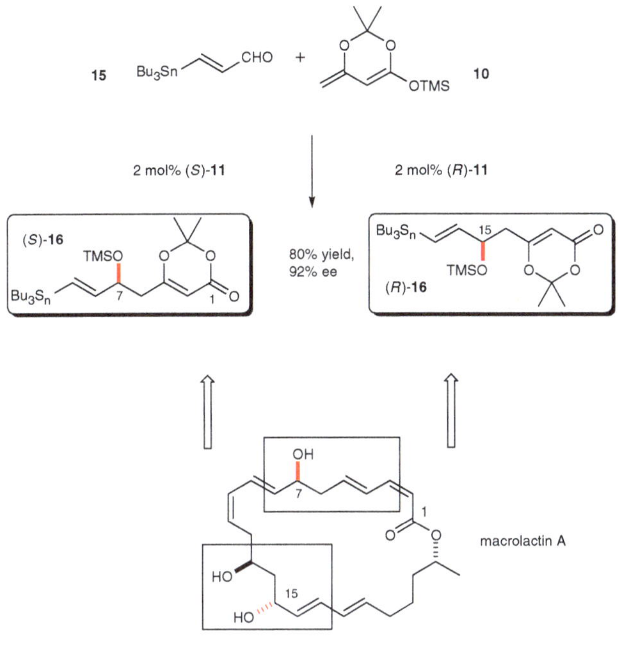

Scheme 3.1.1.10 Total synthesis of macrolactin A

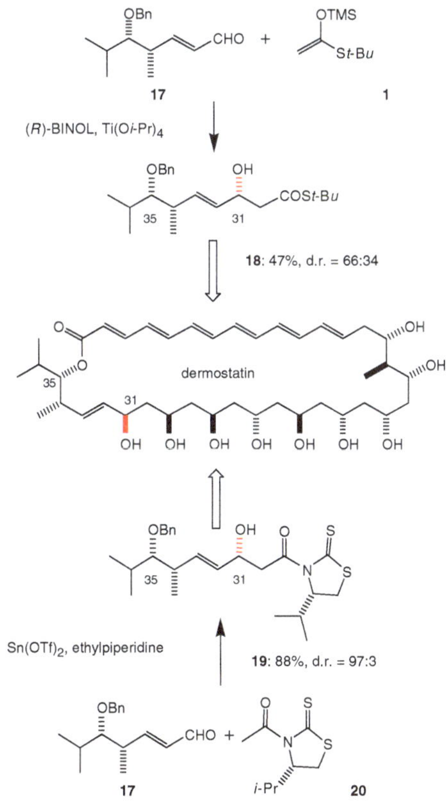

Scheme 3.1.1.11 Total synthesis of dermostatin A – a comparison between titanium Lewis acid-catalysed Mukaiyama and tin enolate aldol addition

Also, the BINOL–titanium(IV) complex (Keck's conditions) was applied in the initial enantioselective vinylogous aldol step in total synthesis of deoxygigantecin. The aldol adduct was obtained with extremely high degree of enantioselectivity (Scheme 3.1.1.12).[49] For an extensive application of $TiCl_4$/EtN*i*-Pr_2 aldol reactions in the total synthesis of deoxygigantecin see Crimmins and She.[50]

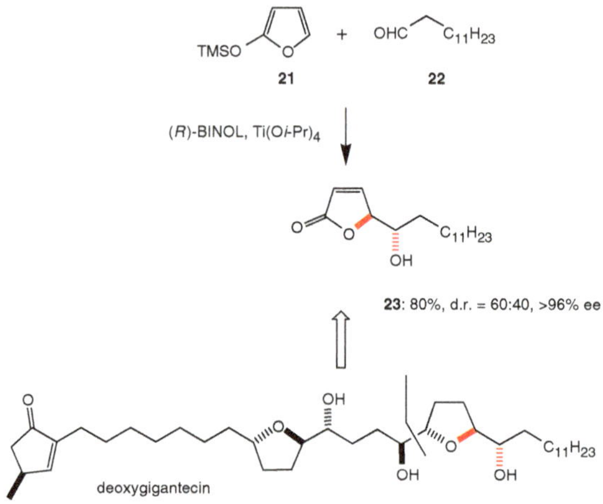

Scheme 3.1.1.12 Total synthesis of (-)-deoxygigantecin

Smith and coworkers used Carreira's catalyst **11** in an acetate aldol step in a highly convergent total synthesis of phorboxazole. When used with 2 mol% of (*R*)-**11** high enantioselectivities and yields of aldol adduct **26** were obtained (Scheme 3.1.1.13).[51] For deployment of $TiCl_4$-mediated Mukaiyama aldol addition in total synthesis of phorboxazole B see Li et al.[52]

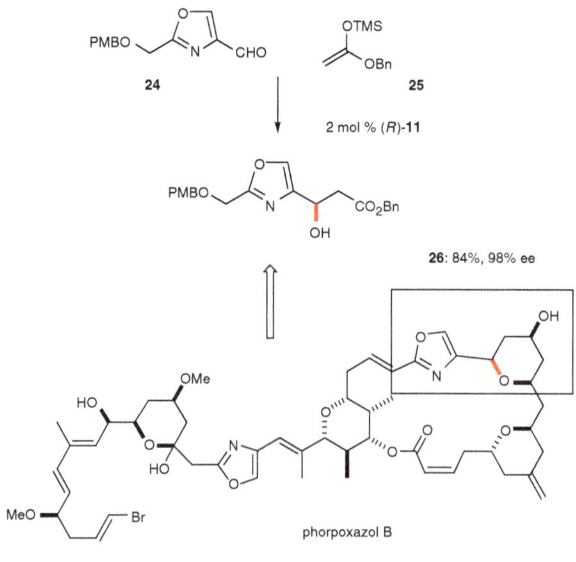

Scheme 3.1.1.13 Total synthesis of phorboxazole B

Further applications of Carreira's catalyst were found in the total synthesis of epinephrine[53] and kedarcidin.[54]

References

1. Mukaiyama T, Banno K, Narasaka K (1974) J Am Chem Soc 96:7503
2. Kitazawa E, Imamura T, Saigo K, Mukaiyama T (1975) Chem Lett 6:569
3. Banno K, Mukaiyama T (1975) Chem Lett 7:741
4. Wong CT, Wong MW (2007) J Org Chem 72:1425
5. Mukaiyama T (1999) Tetrahedron 55:8609
6. Duthaler RO, Hafner A (1992) Chem Rev 92:807
7. Gennari C (1993) In: Comprehensive Organic Synthesis, Trost BM, Fleming I, Heathcock CH (eds). Pergamon, Oxford, vol 2, p 629
8. Braun W (1996) In: Houben-Weyl – Methoden der Organischen Chemie, Helmchen G, Hoffmann RW, Mulzer J, Schaumann E (eds). Thieme, Stuttgart, vol E21b, p 1603
9. Nelson SG (1998) Tetrahedron: Asymm 9:357
10. Carreira EM (2000) In: Lewis Acids in Organic Synthesis, Yamamoto H (ed). Wiley-VCH, Weinheim, vol 1, p 227
11. Carreira EM, Singer RA (1996) Drug Discovery Today 1:145
12. Urabe H, Sato F (2000) In: Lewis Acids in Organic Synthesis, Yamamoto H (ed). Wiley-VCH, Weinheim, vol 2, p 653
13. Mikami K, Terada M (2000) In: Lewis Acids in Organic Synthesis, Yamamoto H (ed). Wiley-VCH, Weinheim, vol 2, p 799
14. Soriente A, De Rosa M, Villano R, Scettri A (2004) Curr Org Chem 8:993
15. Mikami K, Matsumoto Y, Shiono T (2003) In: Science of Synthesis, Bellus D, Jacobsen NE, Ley SV, Noyori R, Regitz M, Reider PJ, Schaumann E, Shinkai I, Thomas EJ, Trost BM (eds). Thieme, Stuttgart, vol 2, p 457
16. Yuan Y, Ding K, Chen G (2008) In: Acid Catalysis in Modern Organic Synthesis. Yamamoto H and Ishihara K (eds) Wiley, Weinheim, vol 2, p 739
17. Reetz MT, Kyong SH, Bolm C, Zierke T (1986) Chem Ind 824
18. Mikami K, Matsukawa S, Kayaki Y, Ikariya T (2000) Tetrahedron Lett 41:1931
19. Zimmer R, Peritz A, Czerwonka R, Schefzig L, Reissig HU (2002) Eur J Org Chem 3419
20. Zimmer R, Schefzig L, Peritz A, Dekaris V, Reissig HU (2004) Synthesis 1439
21. De Rosa M, Dell'Aglio R, Soriente A, Scettri A (1999) Tetrahedron: Asymm 10:3659
22. De Rosa M, Soriente A, Scettri A (2000) Tetrahedron: Asymm 11:3187
23. De Rosa M, Acocella MR, Villano R, Soriente A, Scettri A (2003) Tetrahedron: Asymm 14:2499
24. De Rosa M, Acocella MR, Soriente A, Scettri A (2001) Tetrahedron: Asymm 12:1529
25. De Rosa M, Acocella MR, Villano R, Soriente A, Scettri A (2003) Tetrahedron Lett 44:6087
26. Soriente A, De Rosa M, Villano R, Scettri A (2000) Tetrahedron:Asymm 11:2255
27. Soriente A, De Rosa M, Stanzione M, Villano R, Scettri A (2001)Tetrahedron: Asymm 12:959
28. Heumann LV, Keck GE (2007) Org Lett 9:4275
29. Mikami K, Matsukawa S (1993) J Am Chem Soc 115:7039
30. Ishii A, Kojima J, Mikami K (1999) Org Lett 1:2013
31. Mikami K, Matsukawa S (1994) J Am Chem Soc 116:4077
32. Keck GE, Krishnamurthy D (1995) J Am Chem Soc 117:2363
33. Nelson SG (1998) Tetrahedron: Asymm 9:357
34. Sato M, Sunami S, Sugita Y, Kaneko C (1995) Heterocycles 41:1435

35. (a) Brodmann T, Lorenz M, Schäckel R, Simsek S, Kalesse M (2009) Synlett 174; (b) Kalesse M (2005) Top Curr Chem 244:43; (c) Hosokawa S, Tatsuta K (2008) Mini-Rev Org Chem 5:1; (d) Denmark SE, Heemstra JR Jr, Beutner GL (2005) Angew Chem Int Ed 44:4682; (e) Casiraghi G, Zanardi F, Appendino G, Rassu G (2000) Chem Rev 100:1929
36. Carreira EM, Singer RA, Lee W (1994) J Am Chem Soc 116:8837
37. Schetter B, Mahrwald R (2006) Angew Chem Int Ed 45:7506
38. Singer RA, Carreira EM (1995) J Am Chem Soc 117:12360
39. Ishimaru K, Monda K, Yamamoto Y, Akiba KY (1998) Tetrahedron Lett 54:727
40. Hayashi M, Yoshimoto K, Hirata N, Tanaka K, Oguni N, Harada K, Matsushita A, Kawachi Y, Sasaki H (2001) Isr J Chem 41:241
41. Morohashi N, Hattori T, Yokomakura K, Kabuto C, Miyano S (2002) Tetrahedron Lett 43:7769
42. Kawase T, Takizawa S, Jayaprakash D, Sasai H (2004) Synth Commun 34:4487
43. Schetter B, Stosiek C, Ziemer B, Mahrwald R (2007) Appl Organomet Chem 21:139
44. Schetter B, Mahrwald R (2006) Org Lett 8:281
45. Fan Q, Lin L, Liu J, Huang Y, Feng X (2005) Eur J Org Chem 3542
46. Rychnovsky SD, Khire UR, Yang G (1997) J Am Chem Soc 119:2058
47. Kim Y, Singer RA, Carreira EM (1998) Angew Chem Int Ed Engl 37:1261
48. Sinz CJ, Rychnovsky SD (2002) Tetrahedron 58:6561
49. Szlosek M, Peyrat JF, Chaboche S, Franck X, Hocquemiller R, Figadere B (2000) New J Chem 24:337
50. Crimmins MT, She J (2004) J Am Chem Soc 126:12790
51. (a) Smith, AB III, Verhoest PR, Minbiole KP, Lim JJ (1999) Org Lett 1:909; (b) Smith AB III, Minbiole KP, Verhoest PR, Schelhaas M (2001) J Am Chem Soc 123:10942
52. Li DR, Zhang DH, Sun CY, Zhang JW, Yang L, Chen J, Liu B, Su C, Zhou WS, Lin GQ (2006) Chem Eur J 12:1185
53. Singer RA, Carreira EM (1997) Tetrahedron Lett 38:927
54. Myers AG, Hogan PC, Hurd AR, Goldberg SD (2002) Angew Chem Int Ed 41:1062

3.1.2 Tin Lewis Acids

In their initial experiments Mukaiyama and coworkers found even tin(IV) chloride to be a promising Lewis acid. Tin(IV) chloride proved to be a mild active Lewis acid with good chelation properties.[1,2,3,4,5] Systematic studies concerning stereoselectivities were conducted by Heathcock[6,7,8] and Reetz.[9,10,11,12] For comprehensive, up-to-date overview see Shiina and Fukui.[13] In contrast to the methods described above the stereoselectivities were strongly influenced by oxygen functionalities of the aldehydes used. These results are summarized in Scheme 3.1.2.1.

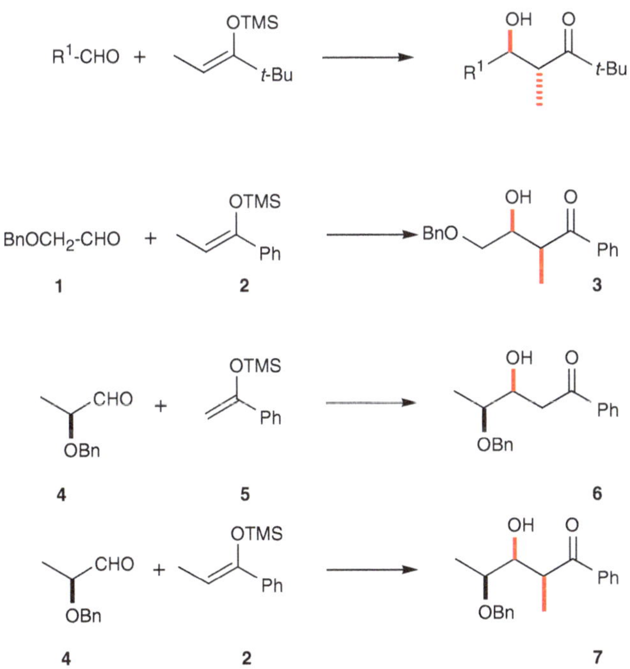

Scheme 3.1.2.1 Dependence of stereochemical outcome on the nature of the aldehydes

Gennari and coworkers extensively investigated the tin(IV) chloride-mediated Mukaiyama reaction of silyl enol ethers of esters as well as thioesters. When used with achiral aliphatic aldehydes poor diastereoselectivities were observed irrespective on configuration of the starting enolates. In contrast to that, high diastereoselectivities were detected in reactions with chiral α-alkoxy aldehydes (Scheme 3.1.2.2).[14,15,16]

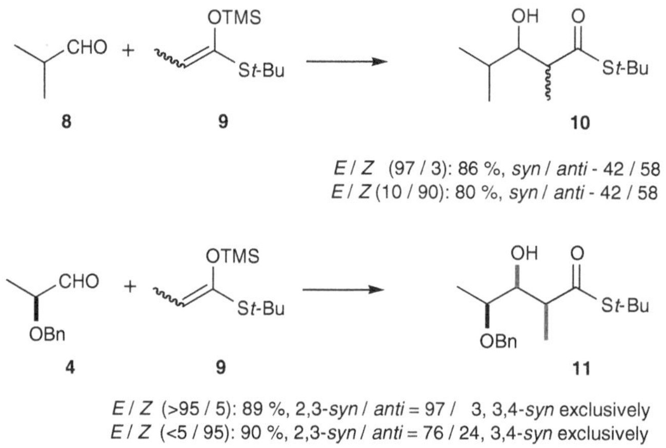

Scheme 3.1.2.2 Comparison of tin(IV) chloride-mediated aldol addition to functionalized and unfunctionalized aldehydes

This elaborated methodology was applied to several total syntheses of natural products, e.g. didemnin,[17,18] tunicamycin[19,20] and bengamide.[21] First successes of enantioselective execution of tin-mediated aldol additions were achieved by the use of chiral tin(II) enolates. This reaction was first reported by Mukaiyama and Iwasawa.[22,23] They successfully tested several chiral diamines in the presence of tin(II) triflate to obtain aldol adducts with a high degree of enantioselectivity.

In a series of publications, Mukaiyama and Kobayashi described the use of stoichiometric amounts of tin(II) triflates and chiral diamines in enantioselective aldol addition of silyl enol ethers with aldehydes.[24,25,26,27,28,29] A complex of dibutyltin acetate, tin(II) triflate and a chiral diamine activate both aldehydes and silylenolates. This 'double activation' provides aldol adducts of aldehydes and silyl enol ethers of carboxylic esters with almost perfect *syn*-diastereo as well enantioselectivities (Scheme 3.1.2.3).

In reactions of enol ethers of α-alkoxyacetic thioesters with aldehydes *syn*- and *anti*-configured enantiopure aldol adducts were synthesized. By the simple exchange of protecting groups of silyl enol ethers *syn*- or *anti*-configured aldol adducts were isolated with high degrees of enantioselectivites (compare results of BnO- with TBSO-substrates in Scheme 3.1.2.4).[30]

Many total syntheses of natural products have been performed by the aid of tin(II) enolate aldol addition, e.g. ribose,[31] L-fucose,[32] leinamycin,[33] rapamycin,[34]

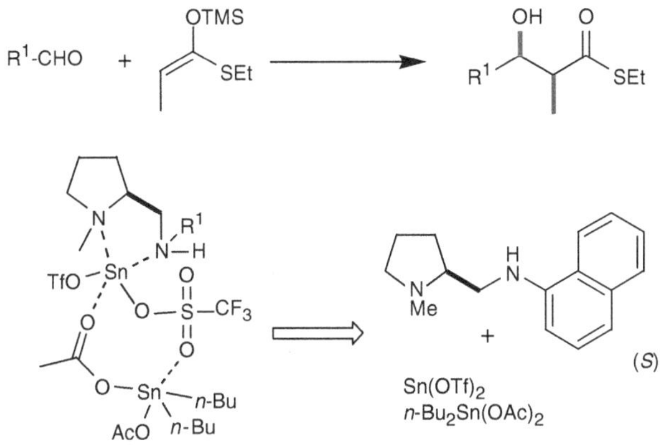

(S)-**12**

entry	R^1	yield (%)	*syn* / *anti*	ee (%)
1	Ph	85	>99 / 1	>98
2	4-ClC_6H_4	96	>99 / 1	>98
3	4-MeC_6H_4	92	>99 / 1	>98
4	4-$MeOC_6H_4$	95	>99 / 1	>98
5	$H_3C(CH_2)_6CHO$	90	>99 / 1	>98
6	Cy	90	>99 / 1	>98
7	*i*-Pr	70	>99 / 1	>98
8	*t*-Bu	86	>99 / 1	>98
9	*trans*-$H_3CHC = CHCH_2$	92	>99 / 1	>98
10	*trans*-PhCH = CH	91	>99 / 1	>98
11	*trans*-*n*-PrCH = CH	91	>99 / 1	>98
12	2-furyl	93	>99 / 1	>98
13	3-thiophenyl	92	>99 / 1	>98

Scheme 3.1.2.3 Mukaiyama reaction with dibutyltin acetate, tin(II) triflate and chiral diamine

sphingosine,[35,36] taxol[37] and references cited in, cephalosporolide D[38,39] and references cited in, octalactin A and B.[40,41,42] A comprehensive overview is given by Shiina.[43]

The methodologies described above represent more or less equimolar versions of tin(II) enolate-mediated aldol additions. To achieve a real catalytic execution one has to keep the concentration of intermediary trimethylsilyl triflate as low as possible in order to prevent the TMSOTf-promoted racemic aldol addition. Following this idea Mukaiyama and Kobayashi developed a protocol for the catalytic execution of these transformations. The starting compounds have to be added slowly to preformed chiral diamine–tin(II) triflate complexes at –78°C (over a period of 3 h).

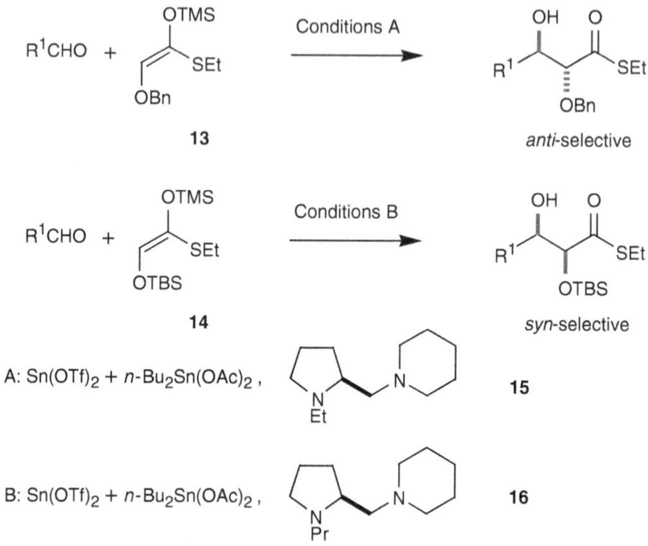

entry	R^1	conditions	yield (%)	*syn* / *anti*	ee % major diastereomer
1	Ph	A	96	1 / 99	96
2	Ph	B	86	88 / 12	90
3	2-furyl	A	83	2 / 98	96
4	2-furyl	B	93	94 / 6	93
5	Et	A	72	2 / 98	97
6	Et	B	46	92 / 8	82
7	*trans*-PhC = CH	A	88	2 / 98	98
8	*trans*-PhC = CH	B	76	90 / 10	92

Scheme 3.1.2.4 Mukaiyama aldol reaction of α-alkoxyacetic acid thioesters

The competitive racemic aldol addition, which arises from high concentrations of intermediary trimethylsilyl triflate, can be suppressed in this way. Thus, aldol adducts were obtained in good yields and high enantioselectivities. Moreover, improvements of selectivities were obtained by the use of propionitrile instead of dichloromethane as solvent (Scheme 3.1.2.5).[44,45]

The selective construction of defined quaternary stereocentres was accomplished by a similar procedure. Kobayashi and coworkers reported the enantioselective Mukaiyama reaction for the synthesis of chiral α-oxygenated carboxylic esters **20** using tin(II) complex with chiral ligand (*S*)-**12** (Scheme 3.1.2.6).[46,47]

A more promising approach concerning the enantioselectivities and catalytic application of tin(II) complexes in Mukaiyama aldol reactions was reported by Evans and coworkers.[48] The authors used their own and well-elaborated

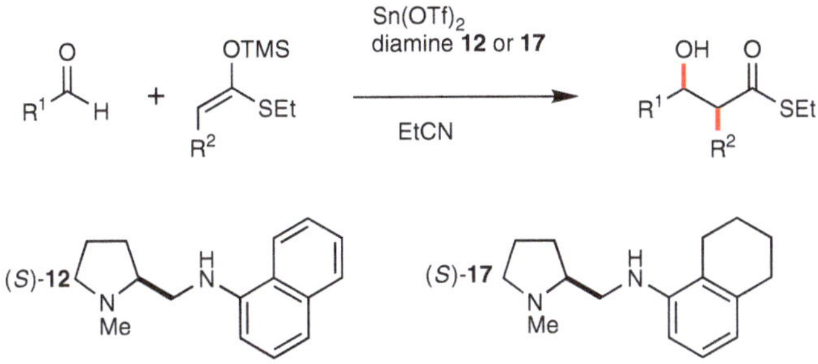

entry	R^1	R^2	diamine	yield (%)	*syn* / *anti*	ee (%)[a]
1	*n*-Bu	H	**17**	79	-	91
2	Cy	H	**17**	81	-	92
3	*i*-Pr	H	**17**	48	-	90
4	PhC≡C	H	**17**	71	-	79[b]
5	Ph	Me	**17**	77	93 / 7	90
6	4-Cl-C_6H_4	Me	**12**	83	87 / 13	90
7	4-Me-C_6H_4	Me	**12**	75	89 / 11	91
8	$H_3C(CH_2)_6$	Me	**12**	80	>99 / 1	>98
9	Cy	Me	**12**	71	>99 / 1	>98

Scheme 3.1.2.5 Catalytic enantioselective execution of the tin(II)-mediated aldol addition
[a]Major diastereoisomer
[b]dichloromethane was used as solvent

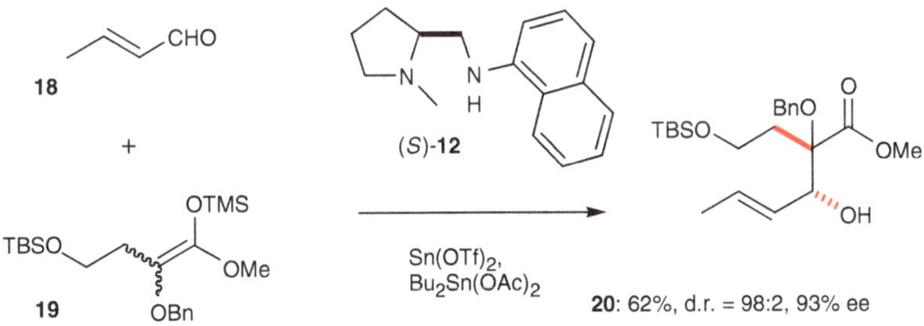

Scheme 3.1.2.6 Synthesis of tertiary alcohols with chiral tin(II) triflates

bis(oxazoline) methodology. Under these conditions silyl enol ether of thioesters reacted with glyoxylates, α-ketoesters and diketones to give the expected chiral tertiary alcohols. Ten mol% of the chiral tin(II) catalysts **24** or **25** were used to afford aldol adducts with high degrees of diastereo- and enantioselectivities (Scheme 3.1.2.7).[49] For an overview of this development see Johnson and Evans.[50]

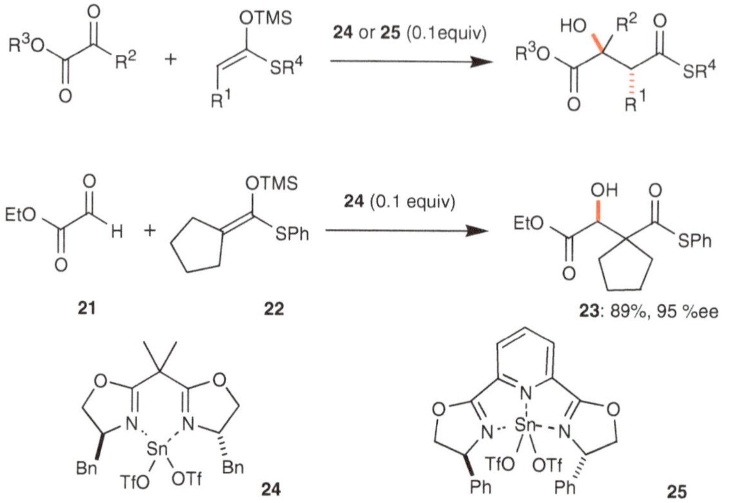

entry	R^1	geometry	R^2	R^3	R^4	catalyst	yield (%)	*syn* / *anti*	ee (%)[a]
1	H	(*Z*)	H	Et	Ph	**24**	90	-	98
2	Me	(*Z*)	H	Et	Ph	**24**	87	90 / 10	95
3	Et	(*Z*)	H	Et	Ph	**24**	90	92 / 8	95
4	*i*-Pr	(*Z*)	H	Et	Ph	**24**	72	93 / 7	95
5	*i*-Bu	(*Z*)	H	Et	Ph	**24**	88	92 / 8	98
6	Me	(*Z*)	Me	Me	*t*-Bu	**25**	94	99 / 1	99
7	Me	(*E*)	Me	Me	*t*-Bu	**25**	84	99 / 1	96
8	Et	(*Z*)	Me	Me	*t*-Bu	**25**	84	98 / 2	97
9	Me	(*Z*)	Me	Me	Et	**25**	91	95 / 5	92
10	Et	(*Z*)	Me	Me	Et	**25**	97	99 / 1	97

Scheme 3.1.2.7 Enantioselective tin-catalysed Mukaiyama reaction of silyl enol ethers of thioesters with glyoxalates, α-ketoesters and α-diketones
[a]Major diastereomer

Kobayashi and coworkers demonstrated the utility of tin Lewis acid-catalysed asymmetric aldol additions in the total synthesis of sphingosines. When used with 20 mol% of chiral ligand (*R*)-**12** optically active intermediate **28** was obtained with high degree of diastereoselectivity (*syn*/*anti* – 97/3) and enantioselectivity (91% ee) (Scheme 3.1.2.8).[51]

Also, this methodology was applied to the total synthesis of phorboxazole B.[52,53] A comparison of these results with those using other metals complexed by pybox-ligands (copper, zinc, etc.) is given by Evans and coworkers in references.[54]

An unexpected and undesirable change to highly *anti*-selective Mukaiyama aldol additions was obtained during the total synthesis of tedanolide 1. The fragment **31** was isolated in nearly 90% yield as one stereoisomer (Scheme 3.1.2.9).[55]

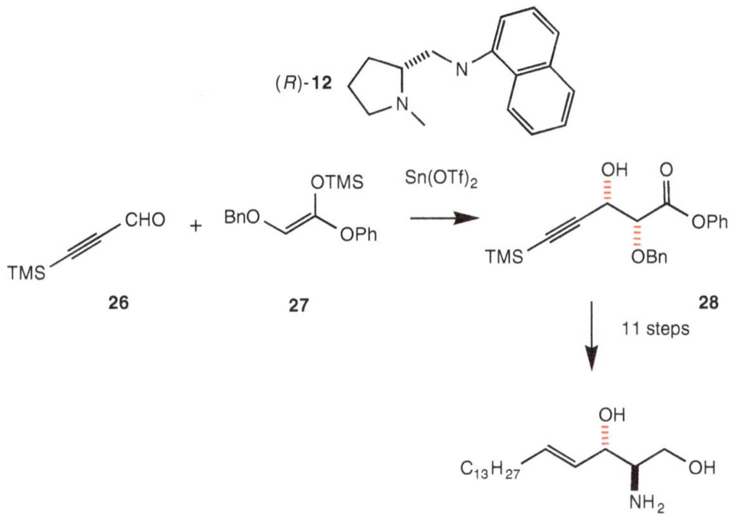

Scheme 3.1.2.8 Total synthesis of sphingosines

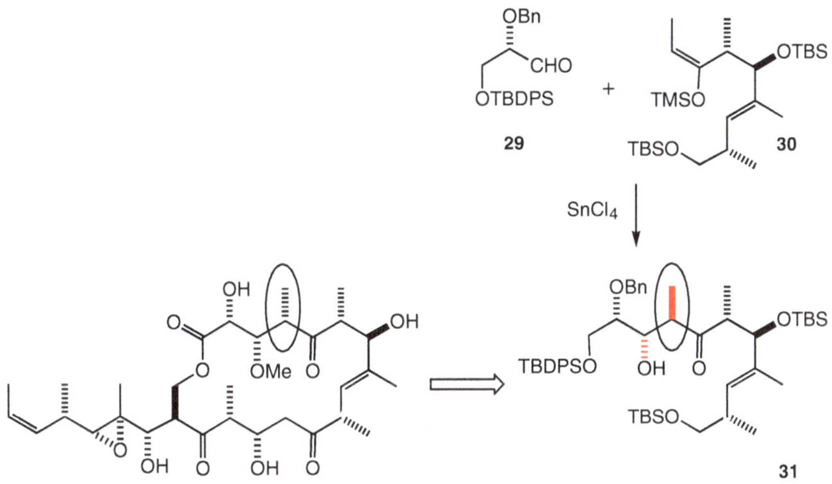

Scheme 3.1.2.9 Unexpected *anti*-Selective $SnCl_4$-catalysed aldol addition in total synthesis of tedanolide

For a $Sn(OTf)_2$-catalysed Mukaiyama reaction in total synthesis of substituted butenolides see Angell et al.[56]

By the $Sn(OTf)_2$-promoted aldol coupling reaction of chiral α-ketoester **33** and silyl ketene thioacetal **32** the important key intermediate **34** was obtained with the required configuration. Thus, the authors were able to demonstrate the simultaneous formation of adjacent quaternary stereocentres at C4 and C5 in a single operation by a $Sn(OTf)_2$-promoted aldol reaction (Scheme 3.1.2.10).[57]

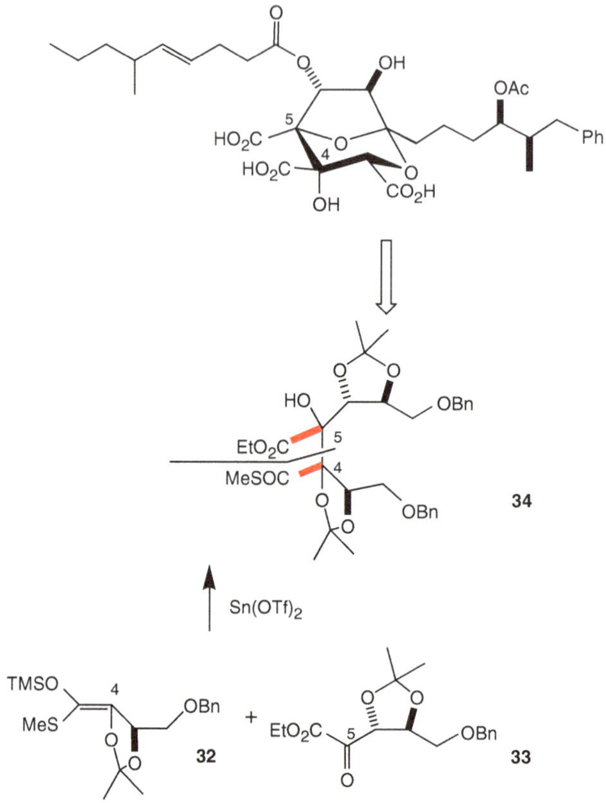

Scheme 3.1.2.10 Tin(II) triflate-catalysed aldol addition in total synthesis of zaragozic acid

In 1997, White and Deerberg published a total synthesis of rapamycin. The polyketide subunit C25–C30 **37** was constructed by an enantioselective Mukaiyama aldol addition using a chiral tin catalyst **38**. High diastereo- as well as enantioselectivities were observed in the reaction of chiral aldehyde **35** with the trimethylsilyl enol ether **36** (*syn*/*anti* – 5/95; 92% ee) (Scheme 3.1.2.11).[58] For an extensive description of the use of boron enolate aldol methodology in total synthesis of rapamycin see also Nicolaou.[59]

Also using chiral tin catalysts, Shiina and coworkers accomplished a total synthesis of octalactin A and B, starting with enantioselective Mukaiyama reactions. All three chiral precursors **39**, **40** and **41** were synthesized by the use of tin(II) triflate and chiral ligands (*S*)-**17**, (*R*)-**12** and (*S*)-**12** with high enantioselectivities (Scheme 3.1.2.12).[60,61] In this way five stereogenic centres were installed selectively. For a review of this work see also Shiina.[62] For an overview of total synthesis of octalactins see also Shiina.[63]

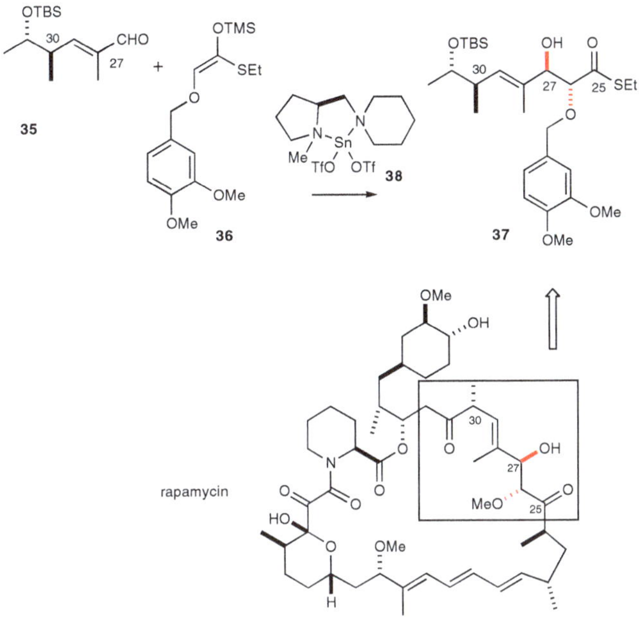

Scheme 3.1.2.11 Total synthesis of rapamycin

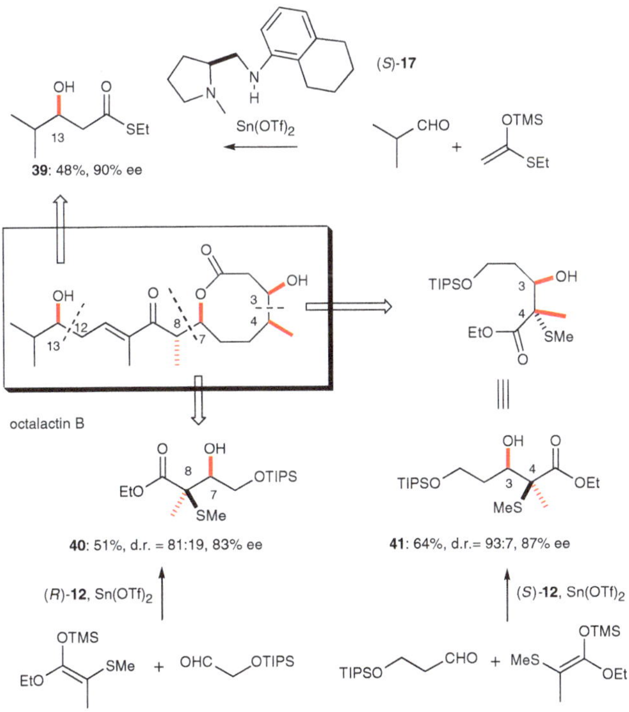

Scheme 3.1.2.12 Total synthesis of octalactin

In a further example Evans and coworkers described an approach to spongistatin 2 using a chiral bis(oxazoline)-tin(II) catalyst. The starting chiral compound **42** was synthesized by means of the (box)–Sn$(OTf)_2$ complexes **43**. The *anti*-configured aldol adduct **42** was obtained with high degrees of enantioselectivity (ee's > 94%). Thus, the C38–C40 region of spongistatin 2 was constructed stereoselectively (Scheme 3.1.2.13)[64] (compare also with Scheme 3.1.4.12 copper-catalysed Mukaiyama reactions).

Scheme 3.1.2.13 Total synthesis of spongistatin 2

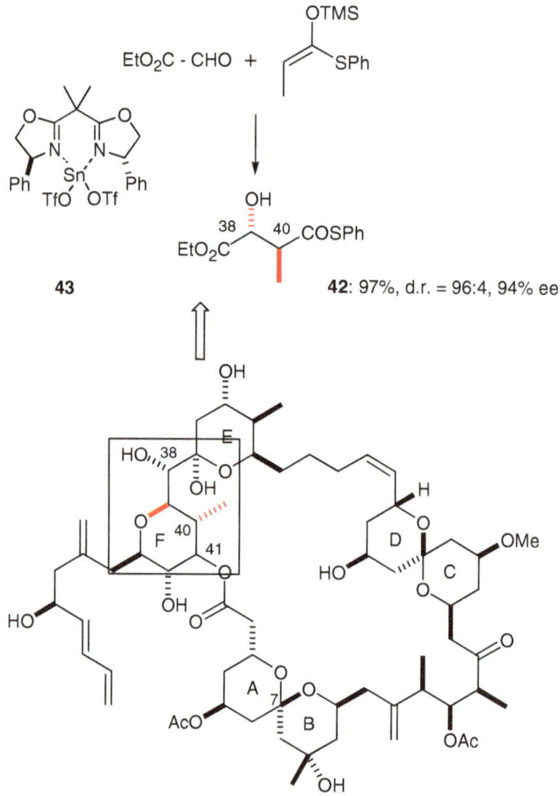

During the total synthesis of phorboxazole the absolute configuration of C15 was established utilizing tin-catalysed asymmetric aldol methodology. By Mukaiyama aldol reactions of aldehyde **44** and silyl ketene thioacetal in the presence of 10 mol% of tin Lewis acid **43** the intermediate **45** was isolated with the required configuration at C15 (Scheme 3.1.2.14).[65] Furthermore, an extensive handling of different aldol methodologies during this total synthesis was reported. By deployment of enantioselective copper-catalysed vinylogous aldol addition the absolute configuration at C5 and C37 was installed. The absolute configuration at C26 and C13 could be established by boron enolate aldol reactions.

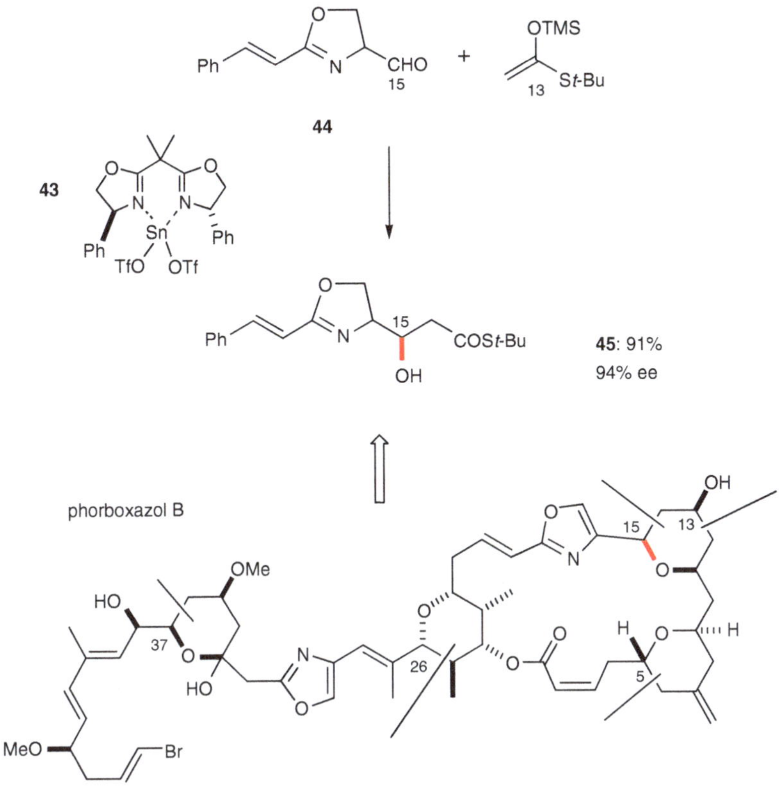

Scheme 3.1.2.14 Total synthesis of phorboxazole B

References

1. Mukaiyama T, Narasaka K, Banno K (1973) Chem Lett 1011
2. Mukaiyama T, Banno K, Narasaka K (1974) J Am Chem Soc 96:7503
3. Saigo K, Osaki M, Mukaiyama T (1975) Chem Lett 989
4. Mukaiyama T (1982) Org React 28:203
5. Mukaiyama T, Narasaka K (1987) Org Synth 65:6
6. Heathcock CH, Hug KT, Flippin LA (1984) Tetrahedron Lett 25:5973
7. Heathcock CH, Davidsen K, Hug KT, Filippin LA (1986) J Org Chem 51:3027
8. Heathcock CH, Montgomery SH (1985) Tetrahedron Lett 26:1001
9. Reetz MT, Kesseler K, Jung A (1984) Tetrahedron Lett 25:729
10. Reetz MT, Kesseler K, Jung A (1984) Tetrahedron 40:4327
11. Reetz MT, Kesseler K, Schmidtberger BW, Steinbach R (1983) Angew Chem Int Ed Engl 22:989
12. Reetz MT, Kesseler K (1985) J Org Chem 50:5434
13. Shiina I, Fukui H (2008) In: Acid Catalysis in Modern Organic Synthesis, Yamamoto H, Ishihara K (eds). Wiley, Weinheim, vol 1, p 537
14. Gennari C, Bernardi A, Poli G, Scolastico C (1985) Tetrahedron Lett 26:2373

15. Gennari C, Beretta MG, Bernardi A, Moro G, Scolastico C, Todeschini R (1986) Tetrahedron 42:893
16. Gennari C, Cozzi PG (1988) Tetrahedron 44:5965
17. Ewing WR, Harris BD, Li WR, Joullié MM (1989) Tetrahedron Lett 30:3757
18. Li WR, Ewing WR, Harris BD, Joullié MM (1990) J Am Chem Soc 112:7659
19. Danishefsky SJ, DeNinno SL, Chen S. Boisvert L, Barbachyn M (1989) J Am Chem Soc 111: 5810
20. Danishefsky S, Barbachyn M (1985) J Am Chem Soc 107:7761
21. Mukai C, Moharram SM, Kataoka O, Hanaoka M (1995) J Chem Soc Perkin Trans 1:2849
22. Mukaiyama T, Iwasawa N, Stevens RW, Haga T (1984) Tetrahedron 40:1381
23. Iwasawa N, Mukaiyama T (1982) Chem Lett 1441
24. Kobayashi S, Mukaiyama T (1989) Chem Lett 297
25. Kobayashi S, Uchiro H, Fujishita Y, Shiina I, Mukaiyama T (1991) J Am Chem Soc 113:4247
26. Mukaiyama T, Kobayashi S (1990) J Organomet Chem 382:39
27. Mukaiyama T, Uchiro H, Shiina I, Kobayashi S (1990) Chem Lett 1019
28. Mukaiyama T, Shiina I, Kobayashi S (1991) Chem Lett 1902
29. Mukaiyama T, Anan H, Shiina I, Kobayahi S (1993) Bull Soc Chim Fr 130:388
30. Mukaiyama T, Shiina I, Uchiro H, Kobayashi S (1994) Bull Chem Soc Jpn 67:1708
31. Mukaiyama T, Shiina I, Kobayashi S (1990) Chem Lett 2201
32. Kobayashi S, Kawasuji T (1993) Synlett 911
33. Kanda Y, Fukuyama T (1993) J Am Chem Soc 115:8451
34. White JD, Deerberg J (1997) Chem Commun 1919
35. Kobayashi S, Furuta T (1998) Tetrahedron 54:10275
36. Kobayashi S, Hayashi T, Kawasuji T (1994) Tetrahedron Lett 35:9573
37. Mukaiyama T, Shiina I, Iwadare H, Saitoh M, Nishimura T, Ohkawa N, Sakoh H, Nishimura K, Tani Y, Hasegawa M, Yamada K, Saitoh K (1999) Chem Eur J 5:121
38. Shiina I, Fujisawa H, Ishii T, Fukuda Y (2000) Heterocycles 52:1105
39. Shiina I, Fukuda Y, Ishii T, Fujisawa H, Mukaiyama T (1998) Chem Lett 831
40. Shiina I, Oshiumi H, Hashizume M, Yamai Y, Ibuka R (2004) Tetrahedron Lett 45:543
41. Shiina I, Kubota M, Oshiumi H, Hashizume M, (2004) J Org Chem 69:822
42. Shiina I, Hashizume M, Yamai YS, Oshiumi H, Shimazaki T, Takasuna YJ, Ibuka R (2005) Chem Eur J 11:6601
43. Shiina I (2004) In Modern Aldol Reactions, Mahrwald R (ed). Wiley-VCH, Weinheim, vol 2, p 105
44. Mukaiyama T, Kobayashi S, Uchiro H, Shiina I (1990) Chem Lett 129
45. Kobayashi S, Fujishita Y, Mukaiyama T (1990) Chem Lett 1455
46. Kobayashi S, Horibe M, Saito Y (1994) Tetrahedron 50:9629
47. Kobayashi S, Fujishita Y, Mukaiyama T (1989) Chem Lett 2069
48. Evans DA, MacMillan DWC, Campos KR (1997) J Am Chem Soc 119:10859
49. Evans DA, Wu J, Masse CE, MacMillan DWC (2002) Org Lett 4:3379
50. Johnson JS, Evans DA (2000) Acc Chem Res 33:325
51. Kobayashi S, Furuta T, Hayashi T, Nishijima M, Hanada K (1998) J Am Chem Soc 120:908
52. Evans DA, Cee VJ, Smith TE, Fitch DM, Cho PS (2000) Angew Chem Int Ed Engl 39:2533
53. Evans DA, Fitch DM (2000) Angew Chem Int Ed Engl 39:2536
54. Evans DA, Wu J, Masse CE, MacMillan DWC (2002) Org Lett 4:3379
55. Jung ME, Zhang, TH (2008) Org Lett 10:137
56. Angell P, Zhang J, Belmont D, Curran T, Davidson JG (2005) Tetrahedron Lett 46:2029
57. Nakamura S, Sato H, Hirata Y, Watanabe N, Hashimoto S (2005) Tetrahedron 61:11078
58. White DJ, Deerberg J (1997) Chem Commun 1919
59. Nicolaou KC (1996) In: Classics in Total Synthesis, Nicolaou KC, Sorensen EJ (eds). Wiley-VCH, Weinheim, p 565
60. Shiina I, Oshiumi H, Hashizume M, Yamai Y, Ibuka R (2004) Tetrahedron Lett 45:543
61. Shiina I, Kubota M, Oshiumi H, Hashizume M (2004) J Org Chem 69:1822

62. Shiina I, Hashizume M, Yamai YS, Oshiumi H, Shimazaki T, Takasuna YJ, Ibuka R (2005) Chem Eur J 11:6601
63. Shiina I (2007) Chem Rev 107:239
64. Evans DA, Trotter BW, Coleman PJ, Cote B, Dias LC, Rajapakse HA, Tyler AN (1999) Tetrahedron 55:8671
65. Evans DA, Fitch, DM, Smith TE, Cee, VJ (2000) J Am Chem Soc 122:10033

3.1.3 Boron Lewis Acids

The classical boron Lewis acids – BF_3 or BCl_3 – were used in stoichiometric amounts in Mukaiyama aldol additions under anhydrous conditions. Though $TiCl_4$ is the more effective Lewis acid in Mukaiyama reactions with aldehydes there are several useful applications of $BF_3 \cdot Et_2O$. BF_3 is known to reverse diastereofacial selectivities in several aldol additions of aldehydes with silyl enol ethers compared with corresponding enolate or Lewis acid-mediated aldol additions. For comprehensive overview of applications of boron Lewis acids in aldol additions see references.[1,2] Much more informations and correlations between substrates and Lewis acids used and stereochemical results are given in Mahrwald.[3] Also in view of this chapter, results of BF_3-mediated aldol additions compared with other Lewis acids used in Mukaiyama aldol reactions were discussed. An explanation for this outstanding behaviour of BF_3 was often given by the non-chelation control of these reactions.

BF_3 was originally suspected to be unable of chelation during aldol additions. Later investigations indicated that the level of 1,2-asymmetric induction in BF_3-mediated aldol additions is also affected by the bulk of the silyl group in the substrate (Scheme 3.1.3.1).[4] In aldol additions of tetrasubstituted silyl enol ether

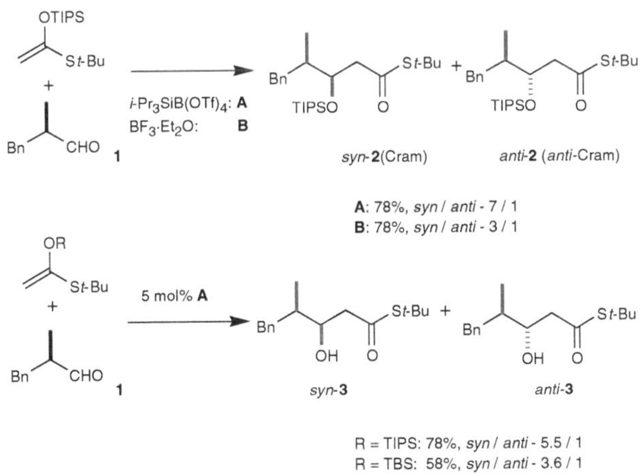

Scheme 3.1.3.1 Selectivities of boron Lewis acid-mediated aldol additions

with oxygen-containing aldehydes a reversal of diastereoselectivity is observed by deployment of Et_2BOTf or BF_3 (Scheme 3.1.3.2).[5]

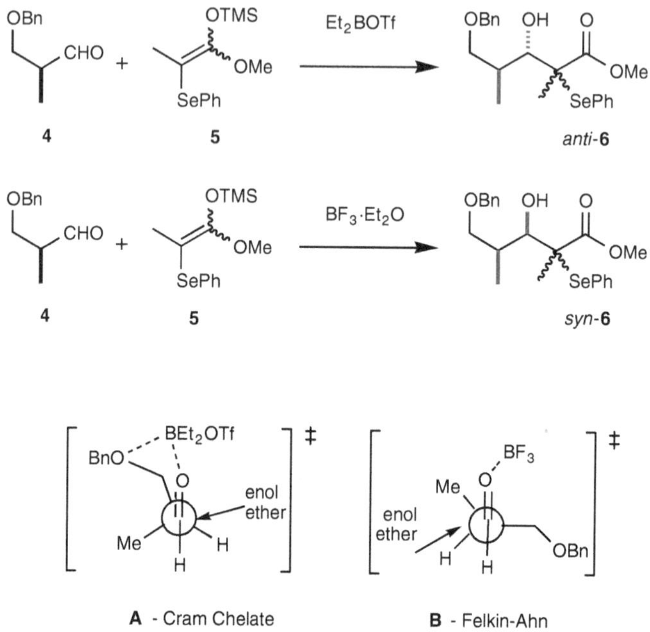

Scheme 3.1.3.2 Diastereoselectivity of boron-mediated aldol additions

For further results of boron Lewis acid-catalysed aldol additions of α-bromo- or α-seleno-substituted silyl ketene acetals see reference.[6] Several other boron Lewis acids were applied in Mukaiyama aldol reactions. Among them are perfluorophenyl borane $B(C_6F_5)_3$[7,8,9] and several diarylborinic acids.[10,11] For a catalytic execution in aqueous media see Mori et al.[12]

Reetz and coworkers reported the first use of a chiral boron Lewis acid in Mukaiyama reactions.[13] The subsequent development of chiral boron Lewis acids has been documented in two comprehensive reviews.[14,15] Later on mainly chiral acyloxyboranes (CAB – boron Lewis acids derived from amino acids) were used in stoichiometric amounts in Mukaiyama aldol additions. Kiyooka and coworkers described the first synthesis and application of these chiral cyclic boranes in 1991.[16] The same group demonstrated the power of this transformation. β-Hydroxyesters were obtained with high degrees of enantioselectivities and yields (Scheme 3.1.3.3).[17,18,19,20] A diastereoselective radical debromination approach coupled with an enantioselective boron Lewis acid-promoted aldol reaction was reported by Kiyooka.[21] When used with this methodology, defined configured stereotriads were obtained.

α-Chiral aldehydes react in these aldol additions by reagent control. Both diastereoisomers are obtained almost in an optically pure form from starting racemic aldehyde. β-Chiral aldehydes react without any Cram selectivity (Scheme 3.1.3.4).[22]

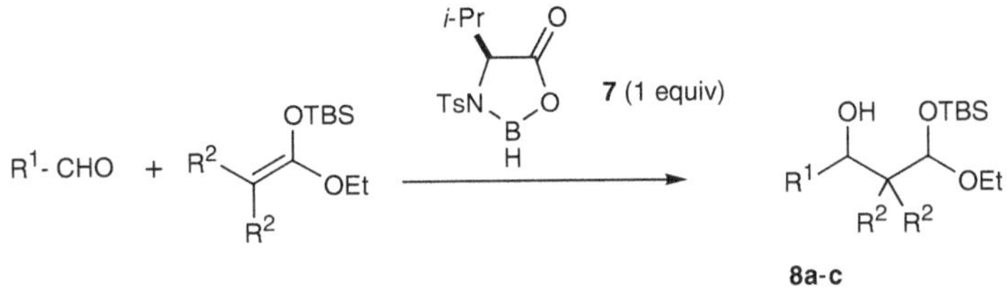

entry	R^1	R^2	yield (%)	ee (%)
1	Ph	H	77	45
2	Ph	Me	83	98
3	*trans*-Ph-CH = CH	H	76	54
4	*trans*-Ph-CH = CH	Me	79	92
5	Ph-$(CH_2)_2$	H	82	62
6	Ph-$(CH_2)_2$	Me	85	96

Scheme 3.1.3.3 Enantioselective aldol reactions of *tert*-butyldimethylsilyl ketene acetals with aldehydes using chiral boron Lewis acid **7**

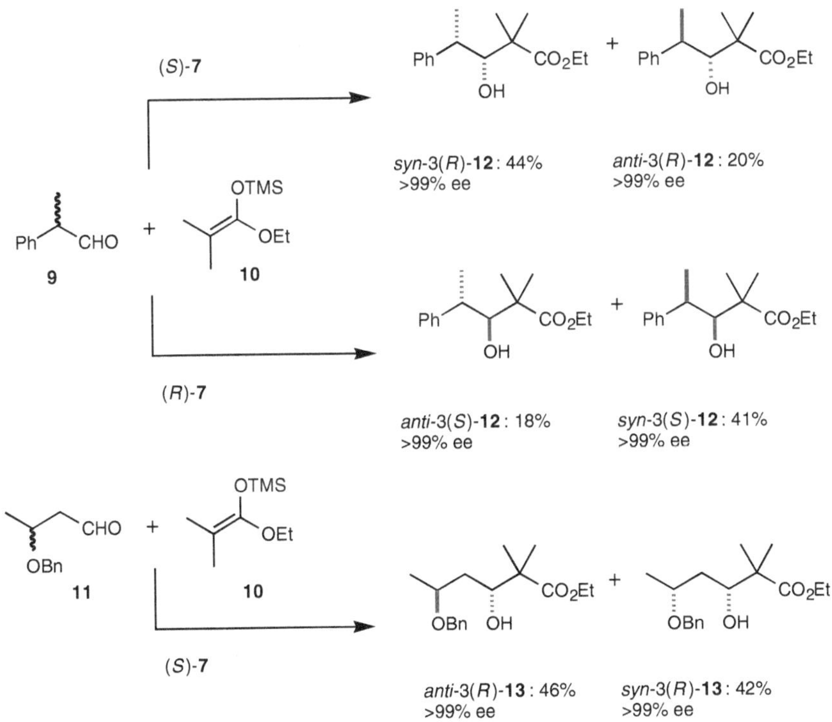

Scheme 3.1.3.4 Aldol additions of α- and β-chiral aldehydes

Applications of this methodology in total syntheses of filipin III[23] and bryostatin[24] were reported by Kyooka and coworkers.

For a catalytic and enantioselective execution of the Mukaiyama reaction several other CAB reagents were developed. Yamamoto and coworkers reported the synthesis and application of a monoacyloxytartaric acid borane. A variety of aldehydes react with enol silyl ethers of ketones in the presence of 10 mol% of chiral acyloxyborane **14a** or **14b** to give β-hydroxyketones in good yields and high enantioselectivities. The aldol adducts were observed with a high degree of *syn*-diastereoselectivity (Scheme 3.1.3.5).[25]

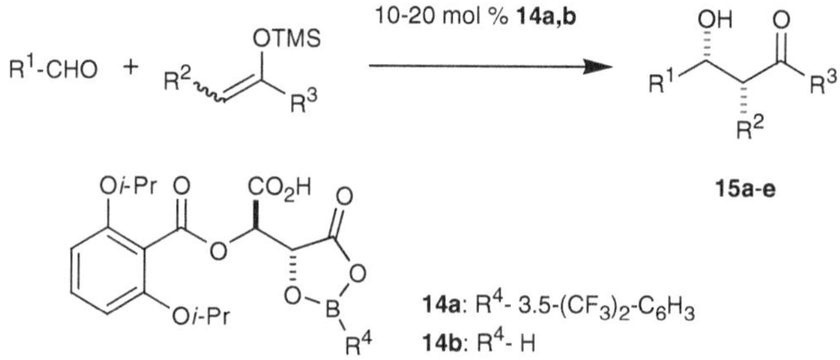

entry	R^1	R^2	R^3	R^4 (mol %)	yield (%)	*syn* / *anti*	ee (%)
1	Ph	H	Ph	**14a** (10)	99	-	88
2	Ph	Me	Ph	**14a** (10)	92	99 / 1	96 (*syn*)
3	Ph	Me	Et	**14b** (20)	99	96 / 4	96 (*syn*)
4	Ph	$-(CH_2)_4-$		**14b** (20)	83	>95 / 5	97 (*syn*)
5	Pr	Me	Et	**14b** (20)	61	80 / 20	88 (*syn*)

Scheme 3.1.3.5 *syn*-Selective enantioselective Mukaiyama aldol addition catalysed by chiral boronic Lewis acid **14a** or **14b**

Moreover, silyl enol ethers of carboxylic esters were reacted with a series of aldehydes in the presence of catalytic amounts of **14b** to give aldol adducts with a high degree of enantioselectivity. Again, aldol adducts were observed with a high degree of *syn*-diastereoselectivity.[26,27]

For an application of this catalyst in vinylogous Mukaiyama aldol additions see Sato et al. (Scheme 3.1.3.6).[28]

The stereoselectivities observed in these reactions can be explained at best by so-called open transition states (Fig. 3.1.3.1).

An AM1-optimized structure of chiral boron Lewis acids was used to develop a transition state model of this reaction (Fig. 3.1.3.2).

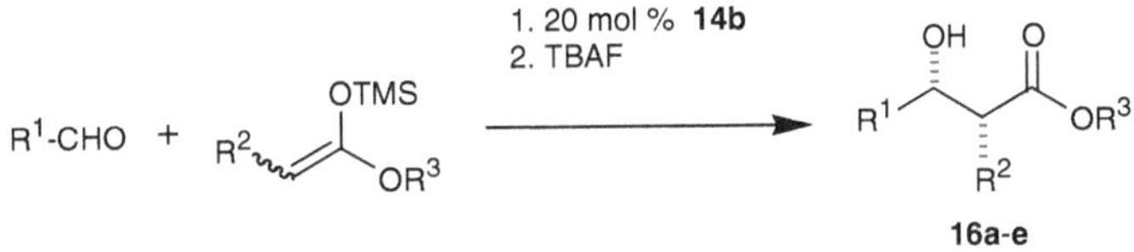

product	R^1	R^2	R^3	*syn / anti*	ee (%)
16a	Ph	H	Ph	-	84
16b	Pr	H	Ph	-	76
16c	Ph	Me	Ph	79 / 21	92 (*syn*)
16d	Pr	Me	Ph	79 / 21	88 (*syn*)
16e	PhCH = CH	Me	Ph	96 / 4	97 (*syn*)

Scheme 3.1.3.6 *syn*-Selective aldol addition of silyl enol ethers of esters

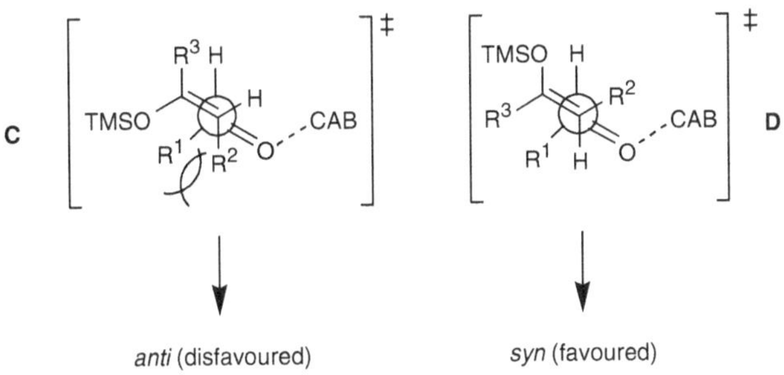

Fig. 3.1.3.1 Extended transition state models

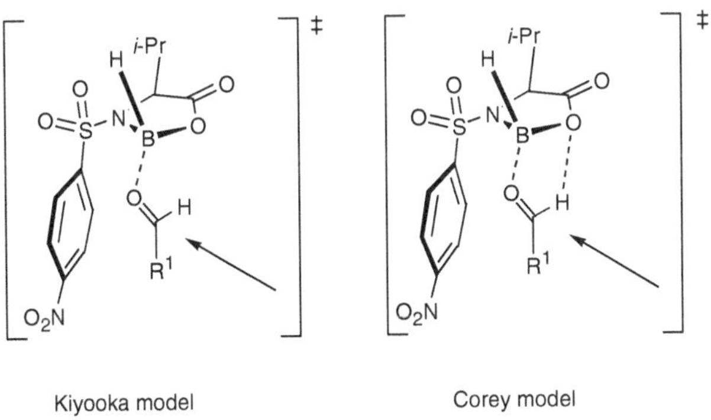

Fig. 3.1.3.2 Proposed transition state models

Later on several other substituted chiral cyclic boranes were developed for the catalytic execution of the Mukaiyama aldol addition. Results of the application of these complexes in reactions with silyl enol ethers of acetophenone are shown in Scheme 3.1.3.7.[29,30]

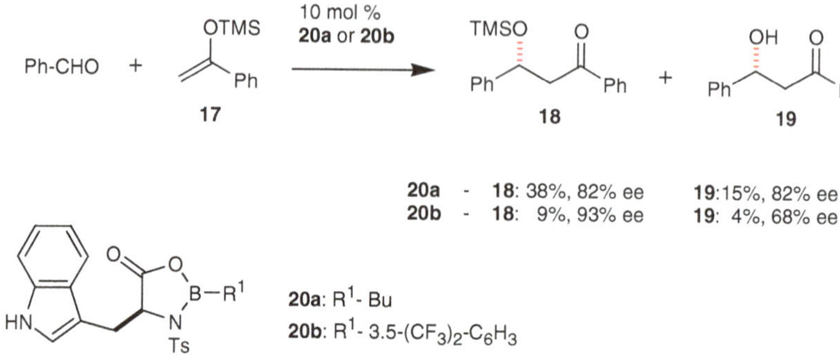

Scheme 3.1.3.7 (*R*)-Selective acetate aldol additions catalysed by boron Lewis acids

For deployment of this boron Lewis acid in vinylogous Mukaiyama aldol reactions see Simsek et al.[31] An application of this boron Lewis acid in vinylogous aldol additions is given in Scheme 3.1.3.8. The aldol adduct **22** was converted upon treatment with mild acidic conditions into the corresponding optical active pyran-4-ones **23**.[32]

For Mukaiyama reactions of silyl enol ethers of substituted ketones see also reference.[32] *Syn*-Configured aldol adducts were obtained as the major diastereoisomers.

Harada and coworkers reported in a series of papers an interesting application of chiral boron Lewis acid-mediated aldol reactions. Cyclic acetals and silyl enol ethers of carboxylic esters were reacted in the presence of 10 mol% of a CAB reagent derived from methyltryptophane. The aldol products were obtained as their corresponding ethers with high degrees of enantioselectivity (Scheme 3.1.3.9).[33,34,35]

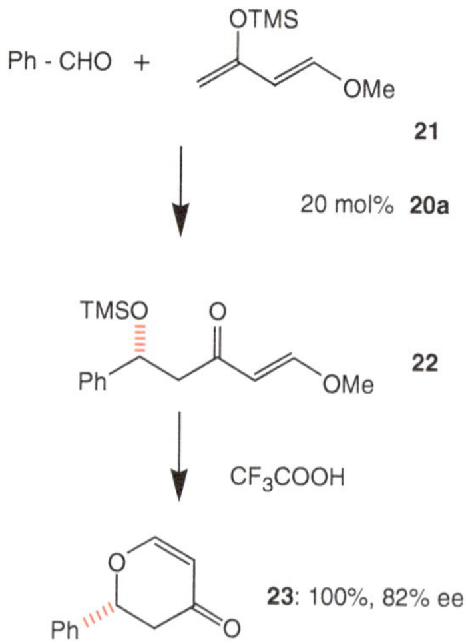

Scheme 3.1.3.8 Synthesis of optically active pyran-4-ones

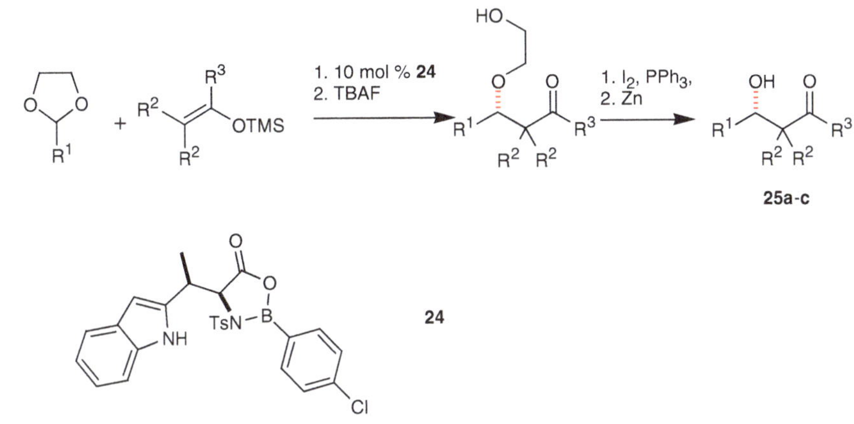

entry	product	R^1	R^2	R^3	yield (%)	ee (%)
1	**25a**	Ph	Me	OEt	88	86
2	**25b**	2-furyl	H	Ph	73	93
3	**25c**	4-MeO-C_6H_4	H	*St*-Bu	80	85

Scheme 3.1.3.9 Reactions of cyclic acetals of aldehydes with silyl enol ether

This methodology has also been used in desymmetrization of meso-1,2-diols. Several CAB reagents were synthesized and tested for this reaction. For results of this investigations see also references[36,37,38,39] (Scheme 3.1.3.10).

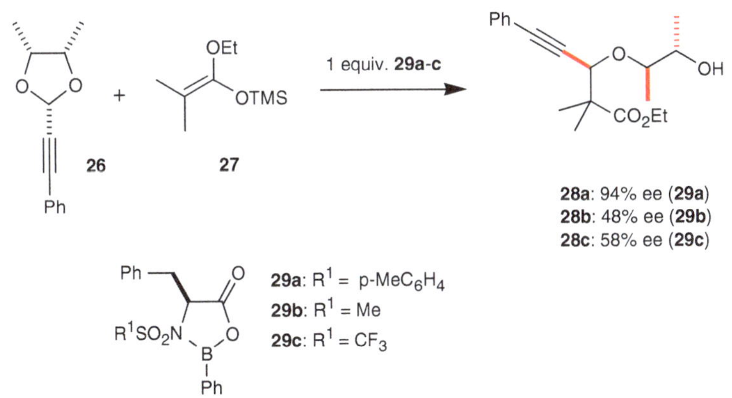

Scheme 3.1.3.10 Desymmetrization of meso-1,2-diols

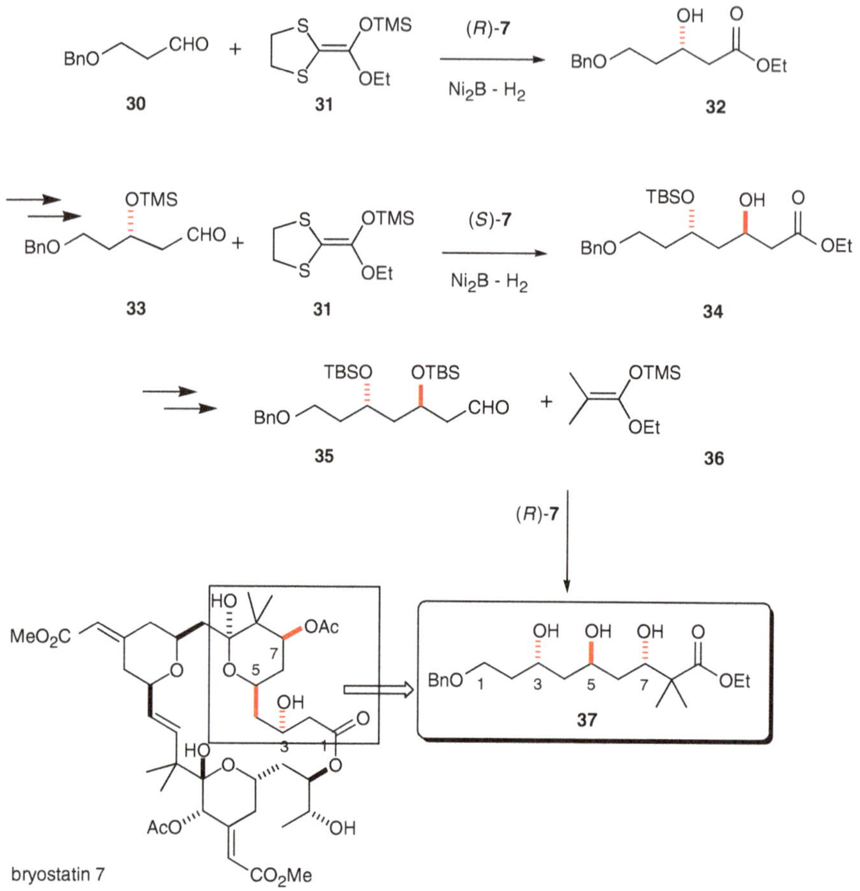

Scheme 3.1.3.11 Total synthesis of bryostatin 7

A very short asymmetric access to a fragment of bryostatin 7, which was accomplished by the use of substoichiometric amounts of chiral boron Lewis acid, was described by Kiyooka and Maeda (Scheme 3.1.3.11).[40] The authors constructed the stereochemical arrangement of triol ester **37** by using three sequential Mukaiyama aldol additions in the presence of chiral oxazaborolidinones (*S*)-**7** and (*R*)-**7** derived from sulfonamides of α-amino acids. For an approach to bryostatin using the boron enolate method see Blanchette et al.[41]

A second example published by Kiyooka et al. demonstrates the power of this aldol method. Filipin III, a polyacetate macrolide, was synthesized by extensive application of chiral oxazaborolidinones (*S*)-**7** and (*R*)-**7** in Mukaiyama aldol reactions (Scheme 3.1.3.12).[42]

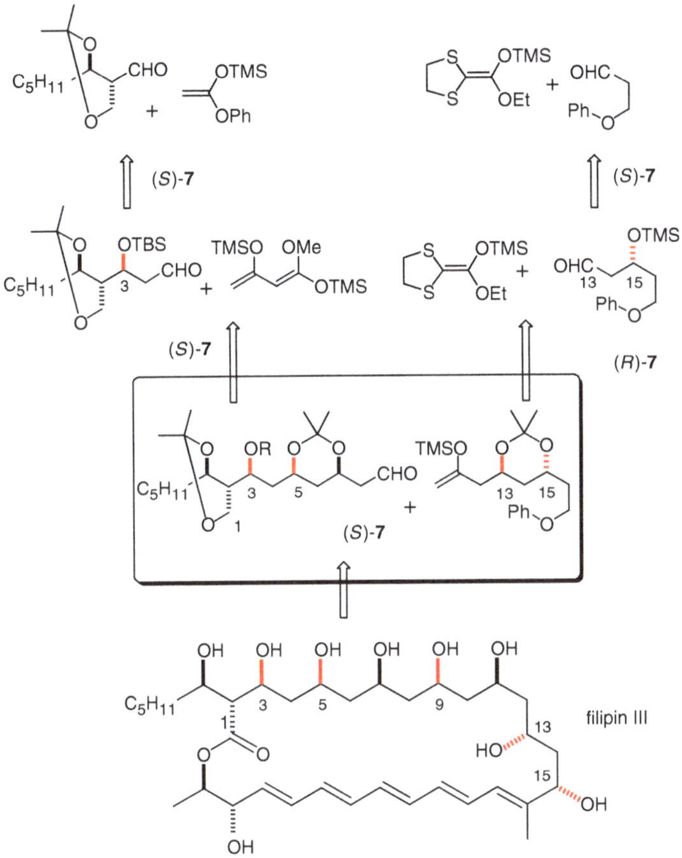

Scheme 3.1.3.12 Total synthesis of filipin III

References

1. Rauniyar V, Hall DG (2008) In: Acid Catalysis in Modern Organic Synthesis, Yamamoto H, Ishihara K (eds). Wiley, Weinheim, vol 1, p 209
2. Ishihara K, Yamamoto H (2004) In: Modern Aldol Reactions, Mahrwald R (ed). Wiley, Weinheim, vol 2, p 25
3. Mahrwald R (1999) Chem Rev 99:1095
4. Davis AP, Plunkett SJ, Muir JE (1998) Chem Commun 1797
5. Guindon Y, Prévost M, Mochirian P, Guérin B (2002) Org Lett 4:1019
6. (a) Guindon Y, Brazeau JF (2004) Org Lett 6:2599; (b) Guindon Y, Houde K, Prévost M, Cardinal-David B, Landry SR, Daoust B, Bencheqroun M, Guérin B (2001) J Am Chem Soc 123:8496
7. Ishihara K, Hanaki N, Yamamoto H (1993) Synlett 577
8. Ishihara K, Funahashi M, Hanaki N, Miyata M, Yamamoto H (1994) Synlett 963
9. Ishihara K, Hanaki N, Funahashi M, Miyata M, Yamamoto H (1995) Bull Chem Soc Jpn 68:1721

10. Ishihara K, Kurihara H, Yamamoto H (1997) Synlett 597
11. Mori Y, Manabe K, Kobayashi S (2001) Angew Chem Int Ed 40:2816
12. Mori Y, Kobayashi J, Manabe K, Kobayashi S (2002) Tetrahedron 58:8263
13. Reetz M, Kunish F, Heitmann P (1986) Tetrahedron Lett 27:4721
14. Deloux L, Srebnik M (1993) Chem Rev 93:763
15. Wallbaum S, Martens J (1992) Tetrahedron: Asymm 3:1475
16. Kiyooka SI, Kaneko Y, Komura M, Matsuo H, Nakano M (1991) J Org Chem 56:2276
17. Kaneko Y, Matsuo T, Kiyooka SI (1994) Tetrahedron Lett 35:4107
18. Kiyooka SI, Kaneko Y, Harada Y, Matsuo T (1995) Tetrahedron Lett 36:2821
19. Kiyooka SI, Hena MA (1996) Tetrahedron: Asymm 7:2181
20. Kiyooka SI, Maeda H (1997)Tetrahedron: Asymm 8:3371
21. Kiyooka SI (2003) Tetrahedron: Asymm 14:2897
22. Kiyooka SI, Kira H, Hena MA (1996) Tetrahedron Lett 37:259
23. Kiyooka SI, Hena MA, Yabukami T, Murai K, Goto F (2000) Tetrahedron Lett 41:7511
24. Kiyooka SI, Maeda H, Hena MA, Uchida M, Kim CS, Horiike M (1998) Tetrahedron Lett 39:8287
25. Furuta K, Maruyama T, Yamamoto H (1991) J Am Chem Soc 113:1041
26. Furuta K; Maruyama T, Yamamoto H (1991) Synlett 439
27. Ishihara K, Maruyama T, Mouri M, Gao Q, Furuta K, Yamamoto H (1993) Bull Chem Soc Jpn 66:3483
28. Sato M, Sunami S, Sugita Y, Kaneko C (1994) Chem Pharm Bull 42:839
29. Ishihara K, Kondo S, Yamamoto H (1999) Synlett 1283
30. Ishihara K, Kondo S, Yamamoto H (2000) J Org Chem 65:9125
31. Simsek S, Horzella M, Kalesse M (2007) Org Lett 9:5637
32. Corey EJ, Cywin CL, Roper TD (1992) Tetrahedron Lett 33:6907
33. Kinugasa M, Harada T, Fujita K, Oku A (1996) Synlett 43
34. Kinugasa M, Harada T, Oku A (1996) J Org Chem 61:6772
35. Kinugasa M, Harada T, Egusa T, Fujita K, Oku A (1996) Bull Chem Soc Jpn 69:3639
36. Kinugasa M, Harada T, Oku A (1997) J Am Chem Soc 119:9067
37. Kinugasa M, Harada T, Oku A (1998) Tetrahedron Lett 39:4529
38. Harada T, Egusa T, Oku A (1998) Tetrahedron Lett 39:5535
39. Harada T, Nakamura T, Kinugasa M, Oku A (1999) Tetrahedron Lett 40:503
40. Kiyooka S, Maeda H (1997) Tetrahedron: Asymm 8:3371
41. Blanchette MA, Malamas MS, Nantz MH, Roberts JC, Somfai R, Whritenour DC, Masamune S (1989) J Org Chem 54:2817
42. Kiyooka S, Hena MH, Yabukami T, Murai K, Goto F (2000) Tetrahedron Lett 41:7511

3.1.4 Copper Lewis Acids

In 1996, chiral (pybox)Cu(II) complexes were employed as catalysts in highly enantioselective aldol additions for the first time. For comprehensive overviews of this development see Stanley and Sibi and others.[1,2,3] The Mukaiyama aldol addition of (benzyloxy)acetaldehyde **2** is catalysed by chiral bis(oxazoline) $Cu(OTf)_2$ **6** and pyridyl bis(oxazoline) Cu(II) complex **7**. Enantioselectivities are significantly lower for aldehydes nominally incapable of chelation (Scheme 3.1.4.1).[4,5]

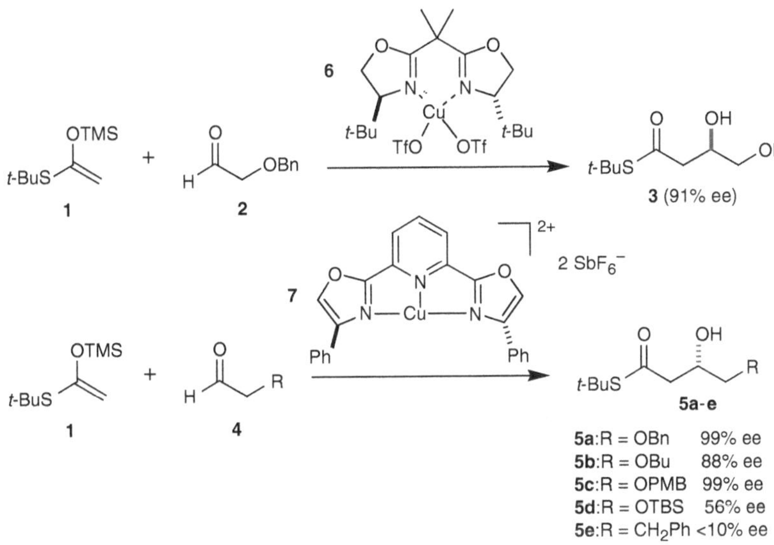

Scheme 3.1.4.1 Addition of enol silanes to α-oxygenated aldehydes

This transformation can be extended to vinylogous substrates. Chan diene **8** and diene acetal **10** react with benzyloxyacetaldehyde to yield the expected δ-hydroxy compounds **9** and **11** with a high degree of enantioselectivity (Scheme 3.1.4.2).

R. Mahrwald, *Aldol Reactions*,

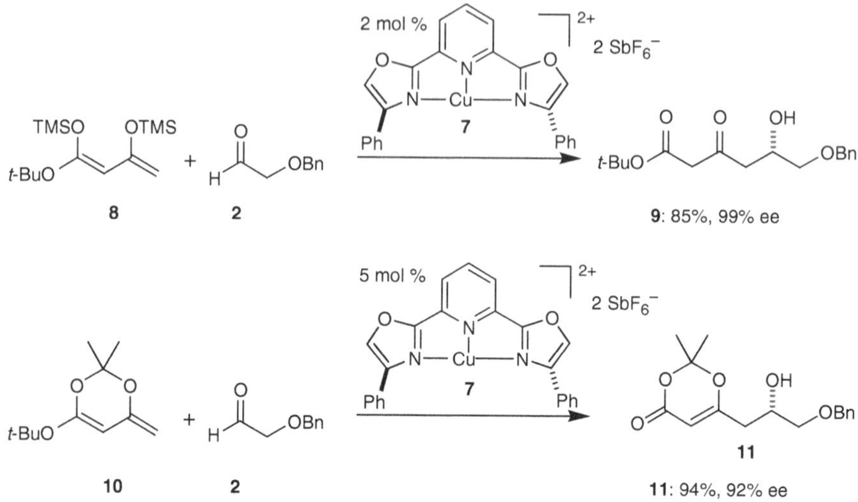

Scheme 3.1.4.2 Enantioselective vinylogous aldol addition

This reaction was successfully employed in the total synthesis of phorboxazole B[6] and bryostatin (Schemes 3.1.4.7).[7] An enantioselective vinylogous aldol addition catalysed by an air-stable hydrated copper catalyst is available for employment in the total synthesis of callipeltoside A (Scheme 3.1.4.8).[8]

Construction of defined quaternary stereocentres can be achieved efficiently by copper-catalysed enantioselective additions of enol silanes to α-ketoesters.

For a direct and catalytic version of copper-catalysed aldol additions of ketoesters see Gathergood et al.[9]

Scheme 3.1.4.3 lists the use of the most effective catalysts with regard to yield and enantioselectivity.[10,11] The hydrated copper complex **14** is an air-stable solid, which is reactive identical to that of complex **6**. The asymmetric pyruvate addition is employed in the enantio- and diastereo-controlled synthesis of α-hydroxy-α-methyl-β-amino acids.[12] Moreover, this transformation can be extended to ketomalonate substrates.[13] In these additions, the tertiary carbinol is not a stereogenic centre.

Kobayashi and coworkers developed an enantioselective aldol addition catalysed by the (*i*-Pr-box)Cu(OTf)$_2$ **17** complex in the presence of water. Under these conditions, a range of enolsilanes undergoes an asymmetric addition to unfunctionalized aromatic, alkenyl and aliphatic aldehydes (Scheme 3.1.4.4). Further experiments have indicated that pure water rather than mixtures of water and organic solvents should be used.[14,15]

Carreira and coworkers developed a highly enantioselective aldol addition using copper salts of BINAP as the chiral source. This approach is used in additions of silyl dienolate **18** to aromatic, heteroaromatic and α,β-unsaturated aldehydes (Scheme 3.1.4.5).[16] The catalyst **20** is generated in situ by treatment of Tol-BINAP, Cu(OTf)$_2$ and (Bu$_4$N)Ph$_3$SiF$_2$. In a mechanistic investigation the authors identified the active species to be a copper(I) enolate.[17]

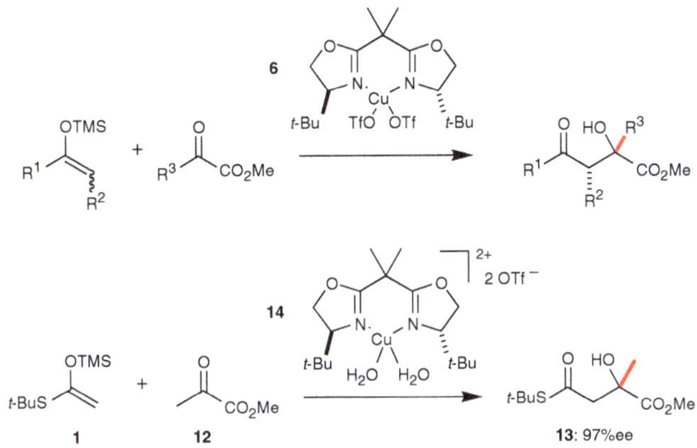

entry	R^1	R^2	R^3	yield (%)	*syn* / *anti*	ee (%) *syn*
1	*t*-BuS	H	Me	95	-	97
2	*t*-BuS	H	Et	84	-	94
3	Ph	H	Me	77	-	99
4	Me	H	Me	81	-	94
5	*t*-BuS	Me	Me	88	97 / 3	99
6	EtS	*i*-Bu	Me	88	90 / 10	93
7	EtS	*i*-Pr	Me	80	90 / 10	99

Scheme 3.1.4.3 Addition of enolsilanes to α-ketoesters

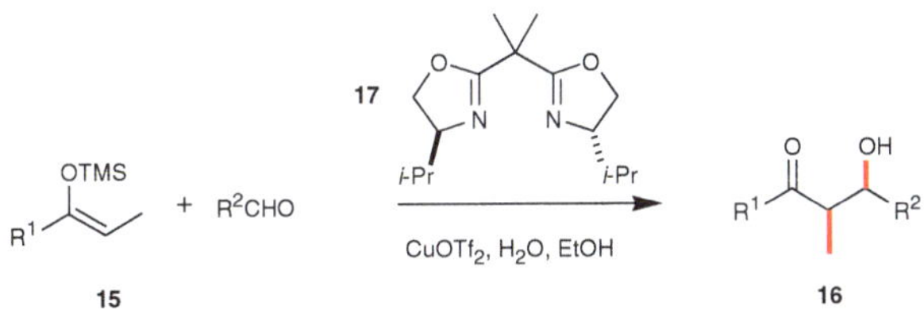

entry	R^1	R^2	yield (%)	*syn*/*anti*	ee (%) *syn*
1	Ph	Ph	74	76 / 24	67
2	Et	Ph	81	78 / 22	81
3	*i*-Pr	Ph	95	80 / 20	77
4	Et	naphthyl	91	80 / 20	79
5	Et	4-Cl-C_6H_4	88	72 / 28	76
6	Et	2-MeO-C_6H_4	87	74 / 26	75
7	*i*-Pr	Cy	77	82 / 18	42
8	*i*-Pr	naphthyl	97	80 / 20	81
9	Ph	4-Cl-C_6H_4	56	61 / 39	67
10	Et	Ph	94	70 / 30	57

Scheme 3.1.4.4 Copper-catalysed aldol addition of silyl enol ethers to unfunctionalized aldehydes

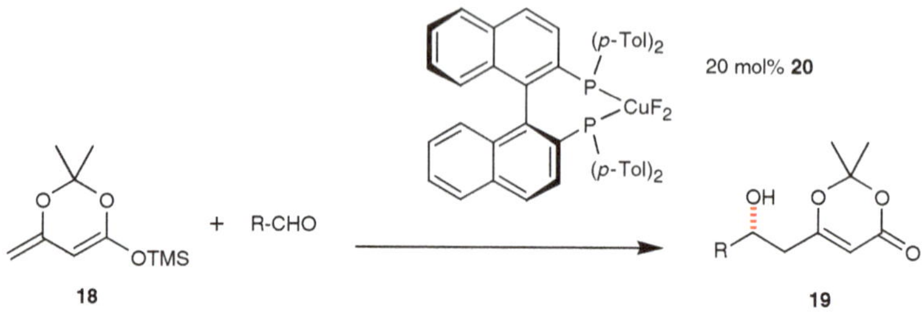

entry	R	yield (%)	ee (%)
1	Ph	92	94
2	naphthyl	86	93
3	C_6H_5-CH = CH-	83	85
4	2-MeO-C_6H_4-CH = CH-	82	90
5	Me_2C = CH-	81	83

Scheme 3.1.4.5 Vinylogous aldol additions of aldehydes

For the use of a CuF–phosphane complex in enantioselective aldol addition of ketones in Mukaiyama reactions see reference.[18] For further developments of new ligand systems and their application in copper-catalysed aldol additions see reference.[19] For a mild and decarboxylative aldol-type addition of malonic acid hemithioesters to aldehydes in the presence of copper catalysts see reference.[20] The highly enantioselective and diastereoselective execution of these aldol reactions of methyl malonic acid half thioester **21** affords *syn*-configured thiopropionic acid aldol adducts.[21] These transformations are compatible with protic functional groups and enolizable aldehydes (Scheme 3.1.4.6).

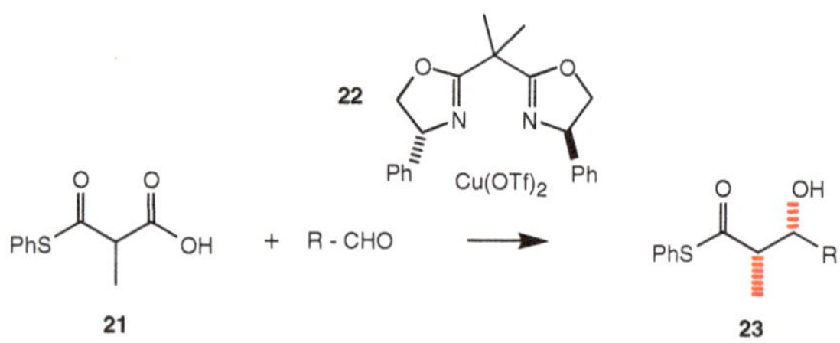

entry	R	yield (%)	*syn* / *anti*	ee (%) (conf.)
1	$Me(CH_2)_6$	80	90 / 10	92 (R)
2	MeO_2C-$(CH_2)_4$	83	91 / 9	94
3	Me-(C = O)-$(CH_2)_8$	71	87 / 13	93
4	Me-$(CH_2)_5$- ≡	59	67 / 33	96 (S)
5	*i*-Bu	73	88 / 12	89
6	Cy	48	97 / 3	93

Scheme 3.1.4.6 Copper-catalysed decarboxylative aldol additions

For investigations of steric influence of ligands in copper-catalysed aldol additions see van Lingen et al.[22] For applications of copper catalysts in vinylogous aldol addition see reference.[23] For copper-catalysed nitro-aldol reaction (Henry-reaction) see reference.[24] For application of polymer-supported chiral copper complexes in Mukaiyama reactions see Orlandi et al.[25]

For the initial generation of chiral starting products in the total synthesis of bryostatin 2 and phorpoxazole B the copper complex (pybox)Cu$(SbF_6)_2$ **7** was deployed. The required starting chiral aldol adduct **25** was obtained with a high degree of enantioselectivity (>99% ee) (Scheme 3.1.4.7).[26]

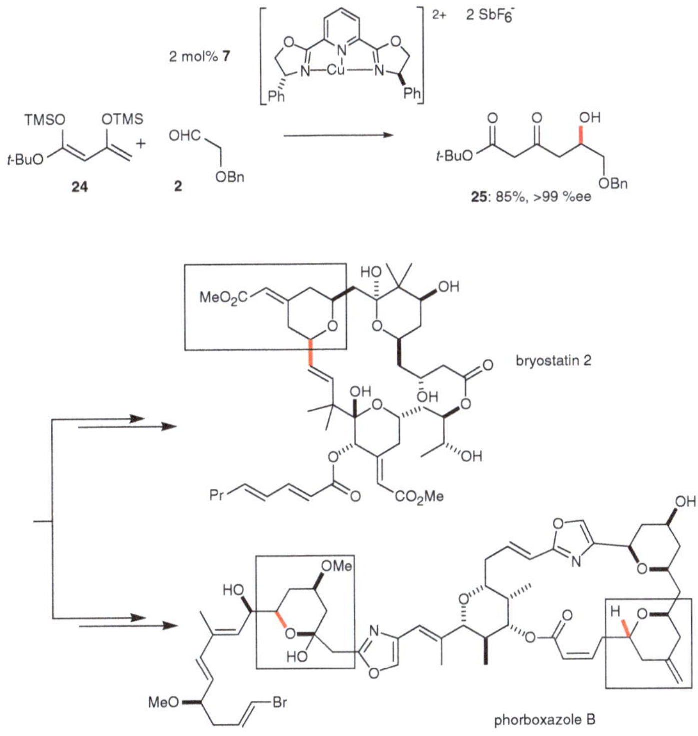

Scheme 3.1.4.7 Total synthesis of phorboxazole B and bryostatin

In the total synthesis of callipeltoside A a vinylogous aldol addition catalysed by the air-stable chiral (Ph-pybox)Cu$(H_2O)_2(SbF_6)$ complex **29** was used for the synthesis of the chiral-substituted acrylic ester **28**. The starting δ-hydroxy-α,β-unsaturated ester **28** was isolated with excellent yield and enantioselectivity (Scheme 3.1.4.8).[27] For total synthesis of callipeltoside using the Paterson boron enolate methodology see reference.[28]

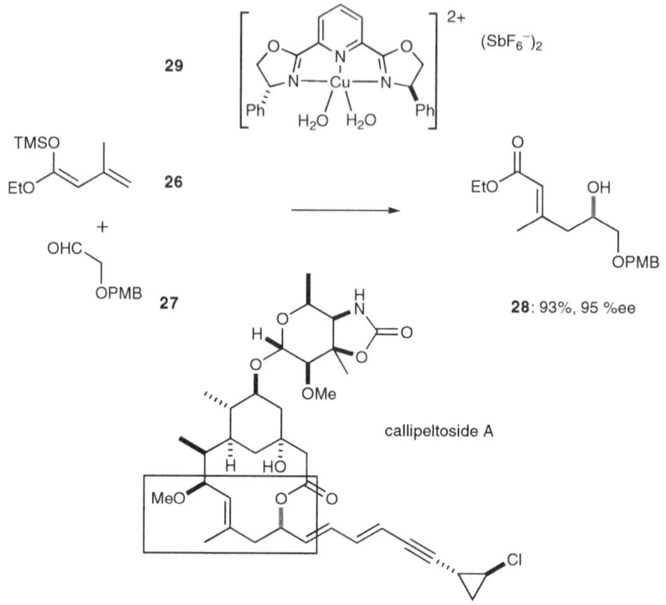

Scheme 3.1.4.8 Total synthesis of callipeltoside A

A representative application of chiral (Tol-BINAP)–CuF_2 complexes (*S*)-**20** and (*R*)-**20** can be found in the total synthesis of amphotericine B (Scheme 3.1.4.9).[29] Both key fragments (*S*)-**32** and (*R*)-**32** of the polyketide subunit C1–C13 were obtained by the same aldol reaction, differing only in the use of (*R*)- or (*S*)-configured (Tol-BINAP)–CuF_2 complex **20**.

A similar approach was used in total synthesis of leucascandrolide A as depicted for amphotericine B (Scheme 3.1.4.9). By the use of 2 mol% of chiral copper(II) triflate the authors obtained the aldol adduct **34** with a high degree of enantiomeric excess by a vinylogous Mukaiyama reaction (Scheme 3.1.4.10).[30]

Campagne and Brennan used the same approach for the starting enantioselective aldol addition in total synthesis of madumycin 1 (Scheme 3.1.4.11).[31] By the application of 10 mol% copper catalyst (*R*)-**20** in a vinylogous aldol addition the authors obtained the chiral intermediate **35** with 81% ee (80% yield).

For application of an initial enantioselective copper-catalysed vinylogous aldol addition in total synthesis of oasomycin A see Evans et al.[32]

Evans and coworkers used in the total synthesis of altohyrtin C a copper-catalysed Mukaiyama acetate aldol addition for the initial installation of the required configuration in the important intermediate **41**. But the extremely high enantioselectivity obtained by copper complex **7**-catalysed aldol reaction was diminished by subsequent Fräter–Seebach α-methylation (dr: 5/1). A comparative study using chiral tin Lewis acids revealed the same high stereoselectivities for aldol additions in the propionate series. Thus, the starting intermediate **40** is available in high yields and enantioselectivities by tin-catalysed aldol addition (Scheme 3.1.4.12).[33]

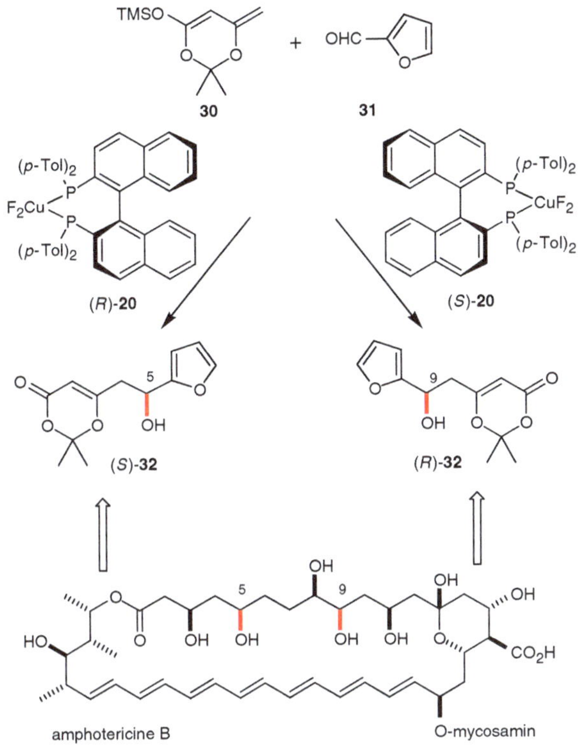

Scheme 3.1.4.9 Total synthesis of amphotericine B

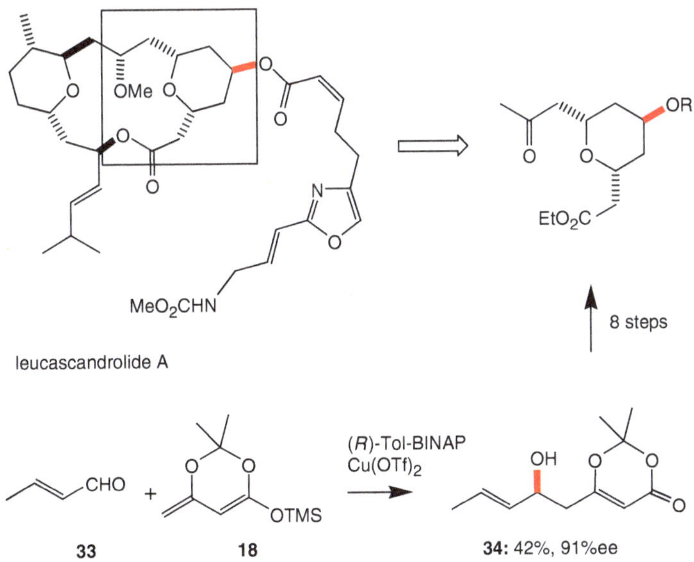

Scheme 3.1.4.10 Total synthesis of leucascandrolide A

Scheme 3.1.4.11 Total synthesis of madumycin 1

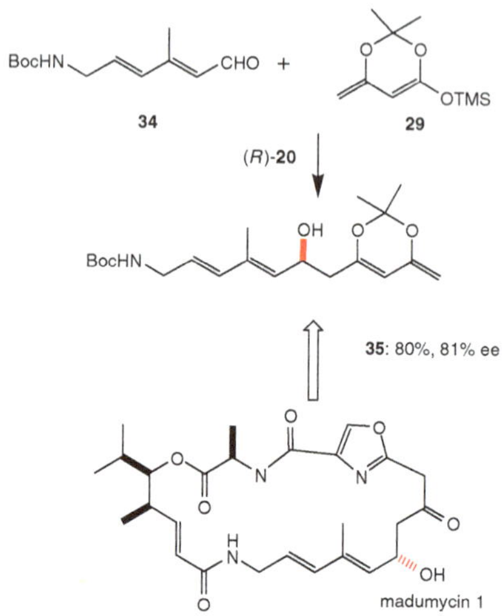

Scheme 3.1.4.12 Total synthesis of altohyrtin – comparison of copper- and tin-catalysed Mukaiyama aldol reaction

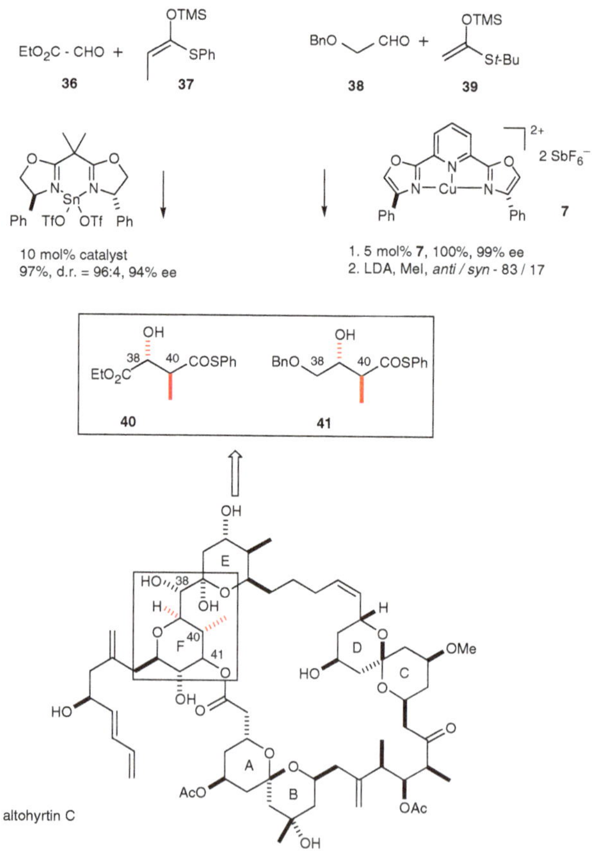

References

1. Stanley LM, Sibi MP (2008) In: Acid Catalysis in Modern Organic Synthesis, Yamamoto H, Ishihara K (eds). Wiley, Weinheim, vol 2, p 911
2. Johnson JS, Nicewicz DA (2004) In: Modern Aldol Reactions, Mahrwald R (ed). Wiley, Weinheim, vol 2, p 69
3. Shibasaki M, Kanai M (2008) Chem Rev 108:2853
4. Evans DA, Murry JA, Kozlowski MC (1996) J Am Chem Soc 118:5814
5. Evans DA, Kozlowski MC, Murry JA, Burgey CS, Campos K, Connell BT, Staples RJ (1999) J Am Chem Soc 121:669
6. Evans DA, Fitch DM, Smith TE, Cee VJ (2000) J Am Chem Soc 122:10033
7. Evans DA, Carter PH, Carreira EM, Charette AB, Prunet JA, Lautens M (1999) J Am Chem Soc 121:7540
8. Evans DA, Hu E, Burch JD, Jaeschke G (2002) J Am Chem Soc 124:5654
9. Gathergood N, Juhl K, Poulsen TB, Thordrup K, Jorgensen KA (2004) Org Biomol Chem 2:1077
10. Evans DA, Kozlowski MC, Burgey CS, MacMillan DWC (1997) J Am Chem Soc 119:7893
11. Evans DA, Burgey CS, Kozlowski MC, Tregay SW (1999) J Am Chem Soc 121:686
12. Roers R, Verdine GL (2001) Tetrahedron Lett 42:3563
13. Reichel F, Fang XM, Yao SL, Ricci M, Jørgensen KA (1999) Chem Commun 1505
14. Kobayashi S, Nagayama S, Busujima T (1998) J Am Chem Soc 120:8287
15. Kobayashi S, Nagayama S, Busujima T (1999) Tetrahedron 55:8739
16. Krüger J, Carreira EM (1998) J Am Chem Soc 120:837
17. Pagenkopf BL, Krüger J, Stojanovic A, Carreira EM (1998) Angew Chem Int Ed 37:3124
18. (a) Oisaki K, Suto Y, Kanai M, Shibasaki M (2003) J Am Chem Soc 125:5644; (b) Suto Y, Kumagai N, Matsunaga S, Kanai M, Shibasaki M (2005) Org Lett 5:3147; (c) Oisaki K, Zhao D, Suto Y, Kanai M, Shibasaki M (2005)Tetrahedron Lett 46:4325; (d) Oisaki K, Zhao D, Kanai M, Shibasaki M (2006) J Am Chem Soc 128:7164; (e) Zhao D, Oisaki K, Kanai M, Shibasaki M (2006) J Am Chem Soc 128:14440
19. (a) Le Julie CD, Pagenkopf BL (2004) Org Lett 6:4097; (b) Benaglia M, Cinquini M, Cozzi F, Celentano G (2004) Org Biomol Chem 2:3401; (c) Simonelli B, Orlandi S, Benaglia M, Pozzi G (2004) Eur J Org Chem 2669; (d) Okamura H, Bolm C (2004) Chem Lett 33:482; (e) Langner M, Bolm C (2004) Angew Chem Int Ed 43:5984; (f) Langner M, Remy P, Bolm C (2005) Chem Eur J 11:6254; (g) Remy P, Langner M, Bolm C (2006) Org Lett 8:1209; (h) Sedelmeier J, Hammerer T, Bolm C (2008) Org Lett 10:917; (i) Langner M, Remy P, Bolm C (2005) Synlett 781; (k) Orlandi S, Benaglia M, Dell'Anna G, Celentano G (2007) J Organomet Chem 692:2120
20. (a) Orlandi S, Benaglia M, Cozzi F (2004) Tetrahedron Lett 45:1747; (b) Lalic G, Aloise AD, Shair MD (2003) J Am Chem Soc 125:2852; (c) Fortner KS, Shair MD (2007) J Am Chem Soc 129:1032
21. Magdziak D, Lalic G, Lee HM, Fortner KC, Aloise AD,MD (2005) J Am Chem Soc 127:7284
22. van Lingen HL, van Delft FL, Storcken RPM, Hekking KFW, Klaassen A, Smits JJM, Ruskowska P, Frelek J, Rutjes FPJT (2005) Eur J Org Chem 4975
23. (a) Bluet G, Campagne JM (2001) J Org Chem 66:4293; (b) Bazan-Tejeda B, Bluet G, Broustal G, Campagne JM (2006) Chem Eur J 12:8358
24. (a) Blay G, Hernandez-Olmos V, Pedro JR (2008) Org Biomol Chem 6:468; (b) Blay G, Climent E, Fernandez I, Hernandez-Olmos V, Pedro JR (2007) Tetrahedron: Asymm 18:1603; (c) Colak M, Aral T, Hosgoeren H, Demirel N (2007) Tetrahedron: Asymm 18:1129; (c) Mansawat W, Saengswang I, U-prasitwong P, Bhanthumnavin W, Vilaivan T (2007) Tetrahedron Lett 48:4235; (d) Bandini M, Benaglia M, Sinisi R, Tommasi S, Umani-Ronchi A (2007) Org Lett 9:2151; (e) Bandini M, Piccinelli F, Tommasi S, Umani-Ronchi A, Ventrici C (2007) Chem Commun 616; (f) Maheswaran H, Prasanth KL, Krishna GG, Ravikumar K, Sridhar B, Kantam ML (2006) Chem Commun 4066; (g) Blay G, Climent E, Fernandez I,

Hernandez-Olmos V, Pedro JR (2006) Tetrahedron: Asymm 17:2046; (h) Gan C, Lai G, Zhang Z, Wang Z, Zhou MM (2006) Tetrahedron: Asymm 17:725
25. Orlandi S, Mandoli A, Pini D, Salvadori P (2001) Angew Chem Int Ed 40:2519
26. Evans DA, Carter PH, Carreira EM, Charette AB, Prunet JA, Lautens M (1999) J Am Chem Soc 121:7540
27. Evans DA, Hu D, Burch JD, Jaeschke G (2002) J Am Chem Soc 12:5654
28. (a) Paterson I, Davies RDM, Marquez R (2001) Angew Chem Int Ed Engl 40:603; (b) Paterson I, Davies RDM, Heimann AC, Marquez R, Meyer A (2003) Org Lett 5:4477
29. Krüger J, Carreira EM (1998) Tetrahedron Lett 39:7013
30. (a) Fettes A, Carreira EM (2002) Angew Chem Int Ed 41:4098; (b) Fettes, A, Carreira EM (2003) J Org Chem 68:9274
31. Brennan CJ, Campagne JM (2001) Tetrahedron Lett 42:5195
32. Evans DA, Nagorny P, Reynolds DJ, McRae KJ (2007) Angew Chem Int Ed 46:541
33. Evans DA, Trotter BW, Coleman PF, Cote B, Dias LC, Rajapakse HA, Tyler AN (1999) Tetrahedron 55:8671

3.1.5 Silver, Palladium and Platinum Lewis Acids

The first asymmetric silver-catalysed Mukaiyama reaction was reported by Yamagishi and coworkers.[1,2] They used a BINOL–$AgPF_6$ complex in aqueous DMF. Yanagisawa and coworkers developed an enantioselective Mukaiyama aldol approach based on the deployment of *p*-Tol-BINAP–AgF. Trimethoxysilyl enol ethers of ketones were reacted with non-enolizable aldehydes. The corresponding *syn*-configured aldol adducts were isolated with excellent enantioselectivities (Scheme 3.1.5.1).[3,4,5]

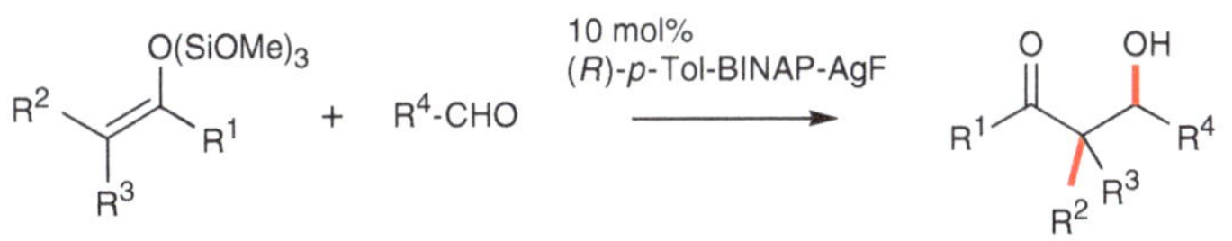

entry	R^1	R^3	R^2	R^4	yield (%)	*syn* / *anti*	ee (%)[a]
1	*t*-Bu	H	Me	Ph	84	>99 / 1	97
2	*t*-Bu	H	Me	4-$MeOC_6H_4$	76	>99 / 1	96
3	*t*-Bu	H	Me	1-naphthyl	63	94 / 6	95
4	-$CH_2(CH_2)_2CH_2$-		H	Ph	78	84 / 16	87
5	-$CH_2(CH_2)_2CH_2$-		H	4-$MeOC_6H_4$	86	75 / 25	92
6	-$CH_2(CH_2)_2CH_2$-		H	4-BrC_6H_4	87	76 / 24	90
7	-$CH_2(CH_2)_2CH_2$-		H	1-naphthyl	68	27 / 73	76
8	-$CH_2(CH_2)_2CH_2$-		H	*trans*-PhCH = CH	81	81 / 19	68
9	-$CH_2CH_2CH_2$-		H	Ph	18	75 / 25	52
10	-$CH_2(CH_2)_3CH_2$-		H	Ph	67	81 / 19	78

Scheme 3.1.5.1 *p*-Tolyl-BINAP–silverfluoride in enantioselective Mukaiyama aldol reactions
[a]Major diastereoisomer

Application of AgF/BINAP complexes in aldol additions of diazo esters were reported by Doyle and coworkers.[6]

An application of silver/BINAP complexes in trichloroacetate-analogous Mukaiyama reactions was described by Yanagisawa and coworkers.[7] High enantioselectivities were detected by applying aromatic aldehydes. For a comprehensive overview of aldol additions catalysed by phosphine/silver complexes see also Yanagisawa and coworkers.[8,9]

Shibasaki and coworkers reported palladium-catalysed enantioselective Mukaiyama reactions.[10,11] Five mol% of BINAP–palladium complex **5** were sufficient for a complete conversion. Good enantioselectivities were obtained by reacting benzaldehyde with enol silyl ethers. Subsequently, the use and application of palladium–aqua complexes **4**[12,13,14] and polymer-supported palladium catalysts in these reactions were described (Scheme 3.1.5.2).[15]

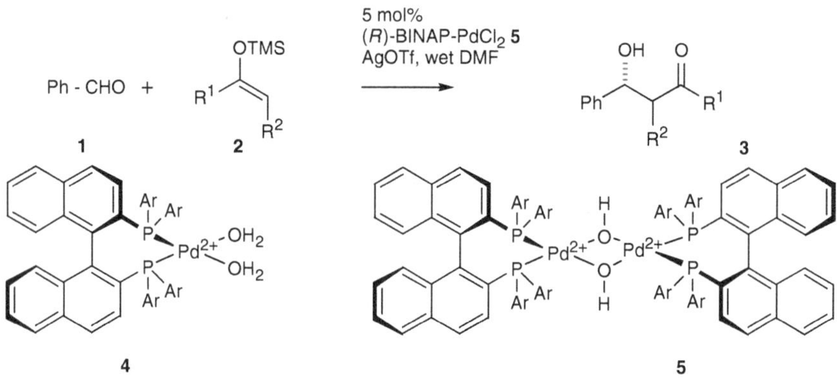

entry	R^1	R^2	yield (%)	*syn* / *anti*	ee (%)[a]
1	Ph	H	87	-	71
2	2-naphthyl	H	80	-	73
3	$-CH_2(CH_2)_2CH_2-$		58	74 / 26	72

Scheme 3.1.5.2 Chiral palladium–BINOL Lewis acids for enantioselective aldol reactions
[a]Major diastereoisomer

For a reinvestigation of this methodology in the presence of silver hexafluoroantimonate see Kiyooka et al.[16]

Fujimura reported the deployment of chiral BINAP-derived platinum complexes in Mukaiyama reactions.[17] When used with 5 mol% of the platinum complexes nearly quantitative yields were obtained. The complexes were activated by the application of equimolar amounts of trifluoroacetic acid. Moderate to high enantioselectivities were detected even when used with enolizable aldehydes (Scheme 3.1.5.3).

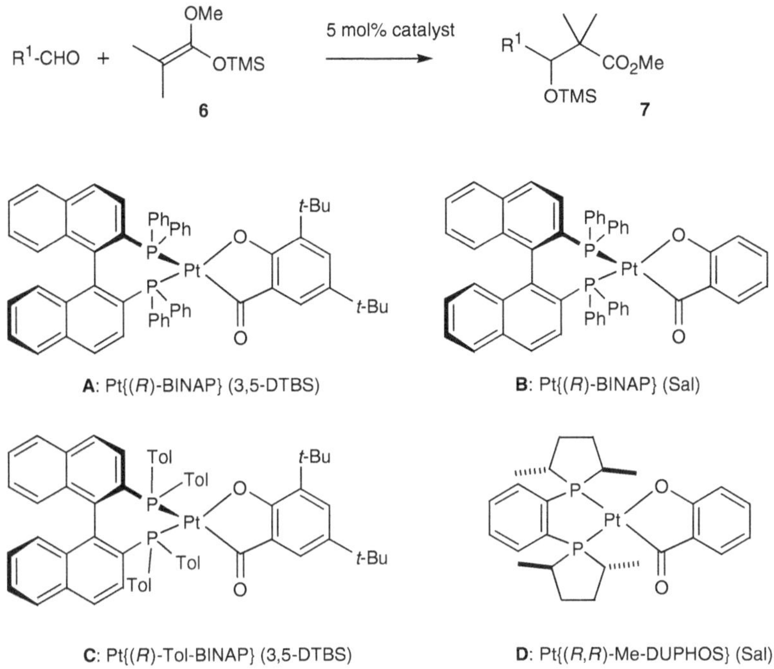

entry	R^1	catalyst	yield (%)	ee (%)
1	Ph	A	99	59
2	Ph	B	99	56
3	Ph	C	97	26
4	Ph	D	99	41
5	*trans*-PhCH = CH	A	90	46
6	*trans*-PhCH = CH	B	92	46
7	$PhCH_2CH_2$	A	94	95
8	$PhCH_2CH_2$	B	96	91
9	*i*-Pr	A	96	88
10	*i*-Pr	B	66	80

Scheme 3.1.5.3 Chiral platinum complexes in enantioselective Mukaiyama aldol reaction

(*R*)-BINAP/platinum complex **8** catalysed the enantioselective aldol reaction even of enolizable aldehydes with 1-methoxy-2-methyl-(1-trimethylsilyloxy)propene **9** at room temperature in dry DMF in high yields and with enantioselectivity up to 92%. This is an example of the versatility of the catalytic enantioselective aldol reaction using a silyl ketene acetal promoted by (μ-hydroxo)–platinum complexes under mild conditions (Scheme 3.1.5.4).[18] For asymmetric adol additions of ketoesters catalysed by chiral dicationic palladium (II) complexes see Mikami and coworkers.[19]

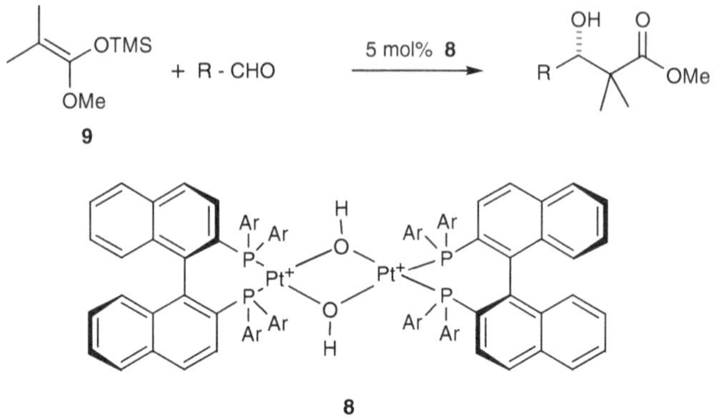

entry	R	yield (%)	ee (%)
1	4-Me-C_6H_4–	82	82
2	4-Ph-C_6H_4	89	82
3	α-naphtyl	92	82
4	*trans*-PhCH = CH	77	72
5	$Ph(CH_2)_2$–	82	92
6	*n*-C_6H_{13}	23	90

Scheme 3.1.5.4 (*R*)-BINAP/platinum catalysed Mukaiyama aldol addition

References

1. Ohkouchi M, Yamaguchi M, Yamagishi T (2000) Enantiomer 5:71
2. Ohkouchi M, Masui D, Yamaguchi M, Yamagishi T (2001) J Mol Catal A: Chem 170:1
3. Yanagisawa A, Nakatsuka Y, Asakawa K, Kageyama H, Yamamoto H (2001) Synlett 69
4. Yanagisawa A, Nakatsuka Y, Asakawa K, Wadamoto M, Kageyama H, Yamamoto H (2001) Bull Chem Soc Jpn 74:1477
5. Wadamoto M, Ozasan N, Yanagisawa A, Yamamoto H (2003) J Org Chem 68:5593
6. Kundu K, Doyle MP (2006) Tetrahedron: Asymm 17:574
7. Yanagisawa A, IIchikawa T, Arai T (2007) J Organomet Chem 692:550
8. Yanagisawa A, Arasi T (2008) Chem Commun 1165
9. Yanagisawa A (2008) In: Acid Catalysis in Modern Organic Synthesis, Yamamoto H, Ishihare K (eds). Wiley, Weinheim, vol 2, p 992
10. Sodeoka M, Ohrai K, Shibasaki M (1995) J Org Chem 60:26
11. Sodeoka M, Shibasaki M (1998) Pure Appl Chem 70:411
12. Sodeoka M, Tokunoh R, Miyazaki F, Hagiware E, Shibasaki M (1997) Synlett 463
13. Kiyooka SI, Hosokawa S, Tsukasa S (2006) Tetrahedron Lett 47:3959
14. Kiyooka SI, Takeshita Y, Tanaka Y, Higaki T, Wada Y (2006) Tetrahedron Lett 47:4453
15. Fujii A, Sodeoka M (1999) Tetrahedron Lett 40:8011
16. Kiyooka SI, Hosokawa S, Tsukasa S (2006) Tetrahedron Lett 47:3959
17. Fujimura O (1998) J Am Chem Soc 120:10032
18. Kiyooka SI, Matsumoto S, Kojima M, Sakonaka K, Maeda H (2008) Tetrahedron Lett 49:1589
19. Mikami K, Kawakami Y, Akiyama K, Aikawa K (2007) J Am Chem Soc 129:12950

3.1.6 Zirconium Lewis Acids

Zirconium compounds were known to be used mostly for forming zirconium enolates by transmetallation for a subsequent aldol addition.[1,2,3,4,5,6,7] Preformed lithium enolates were transmetallated to yield predominantly (*Z*)- or (*E*)-zirconium enolates. Irrespective of the geometry of the zirconium enolates used a preference to *syn*-configured aldol adducts is observed. This methodology was also deployed in highly diastereoselective aldol reactions to obtain chiral aldol adducts (Scheme 3.1.6.1).[8,9,10,11,12,13,14]

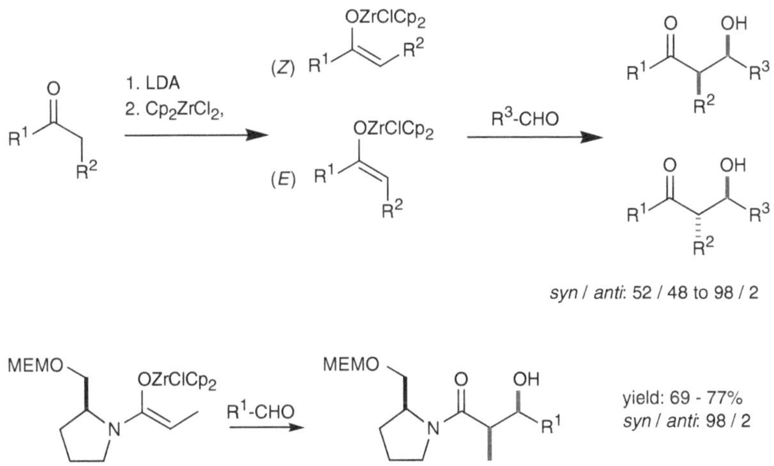

Scheme 3.1.6.1 Stereoselective aldol reactions using zirconium enolates

In addition to that, zirconium(IV) alkoxides are able to deprotonate α-carbon atoms of ketones.[15,16] By subsequent reactions with aldehydes the corresponding aldol adducts were formed.[17] The regioselectivity of this aldol addition is illustrated in Scheme 3.1.6.2.

Even in Mukaiyama reactions zirconium(IV)-alkoxides act as mild Lewis acids.[18,19,20,21] For an overview see Kanno et al.[22] Zirconium(IV)-alkoxides react with aldehydes and silyl enol ethers of carboxylic esters or thioesters for an efficient access to the expected aldol adducts with a high degree of stereoselectivity

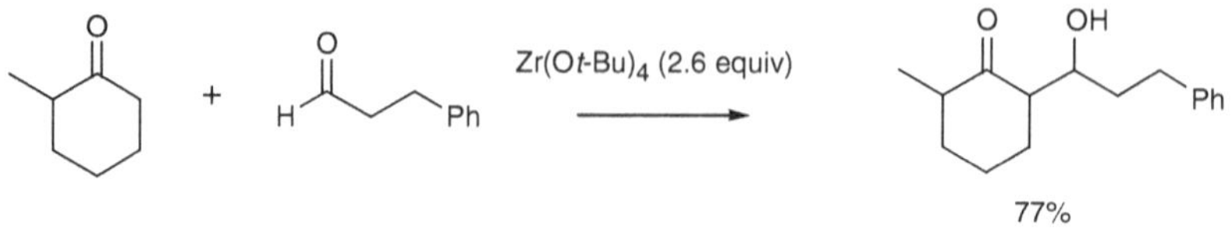

1 (2.0 equiv) 2 3

Scheme 3.1.6.2 Direct aldol reaction using zirconium(IV) alkoxides

(Scheme 3.1.6.3). Kobayashi and coworkers developed a catalytic and enantioselective version of the Mukaiyama reaction based on the use of chiral 3,3'-I_2–BINOL–zirconium Lewis acids (Fig. 3.1.6.1).[23,24] Independent of the geometry of the starting keten silyl acetals *anti*-configured aldol adducts were obtained with high degrees of selectivity. This is true for both diastereo- as well as enantioselectivities.

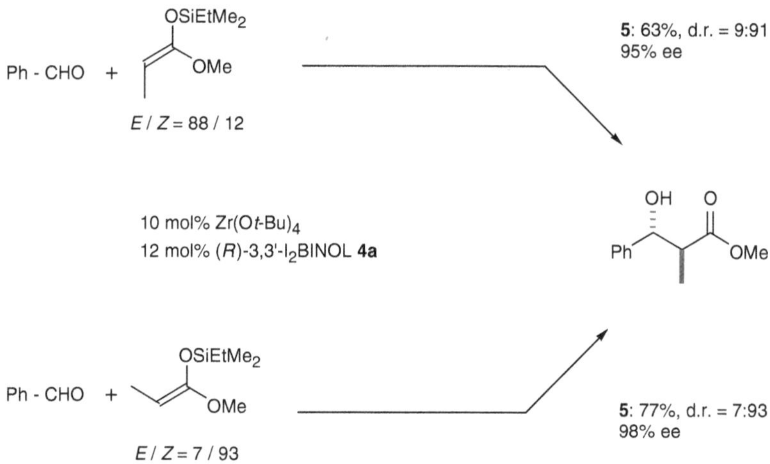

Scheme 3.1.6.3 Influence of geometry of silyl enol ethers

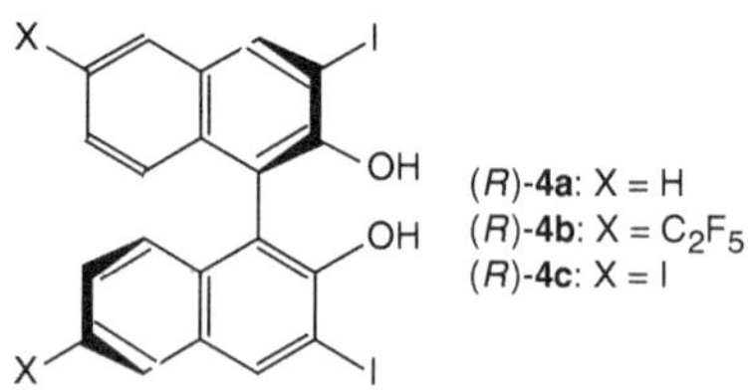

Fig. 3.1.6.1 Chiral BINOL ligands

This general *anti*-preference is a salient feature of zirconium-mediated Mukaiyama aldol reactions. For recent examples of *anti*-selective aldol reactions see references.[25,26,27,28,29,30,31,32,33,34] It was found that additional amounts of alcohol were crucial. For the effects of additional alcohol in catalytic reactions see also Evans and Johnson and others.[35,36,37] The same is true for addition of water to the reaction mixture. Exclusion of water resulted in noticeably lower enantioselectivities. For similar observations made in metal-catalysed reactions see also Posner et al. and others.[38,39,40,41] Results of this development in the enolizable aldehyde series are shown in Scheme 3.1.6.4.

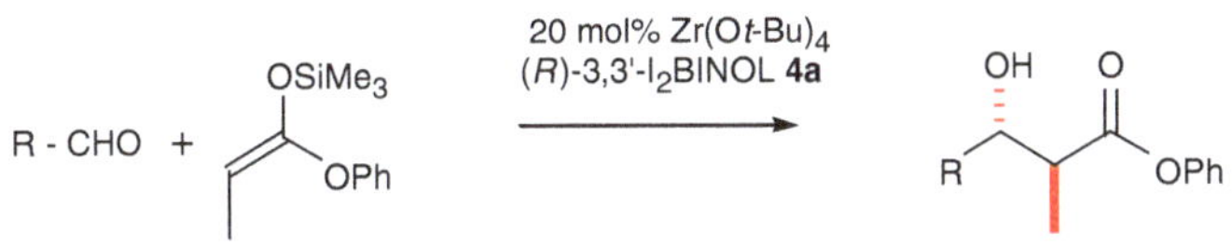

6a-h

entry	R	yield (%)	*syn* / *anti*	ee (%)
1	C_5H_{11}	64	12 / 88	85
2	Ph-$(CH_2)_2$–	71	10 / 90	82
3	C_3H_7	71	15 / 85	81
4	*i*-Pr-$(CH_2)_2$–	56	12 / 88	89
5	Cy-$(CH_2)_2$–	52	14 / 86	78
6	*i*-Pr-CH_2–	16	17 / 83	28
7	Cy-CH_2–	4	21 / 79	31
8	Cy-	traces	-	-

Scheme 3.1.6.4 Influence of structure of aliphatic aldehydes in zirconium-catalysed Mukaiyama reactions

Subsequently, the same group developed an air-stable chiral zirconium Lewis acid for the Mukaiyama reaction of silyl enol ethers of carboxylic esters. This catalyst – ZrMS – is storeable for 4 months. Yields and enantioselectivities remain unaffected. The catalyst was prepared by simply combining zirconium propoxide and BINOL-**4a** and molsieve MS 5A containing 10% water.[42] Some results of this development are shown in Scheme 3.1.6.5.

A very early application of zirconium enolate aldol addition was reported by Deslongchamps and coworkers in total synthesis of erythronolide A.[43,44]

Corey and coworkers used zirconium enolates of corresponding propionate-HYTRA in a stereoselective aldol step during the total synthesis of lactacystin (Scheme 3.1.6.6).[45]

Chiral zirconium catalyst **4a** was deployed in total synthesis of optically active sphingosine. The optically active intermediate **10** was isolated with the required configuration in high yields (Scheme 3.1.6.7).[46]

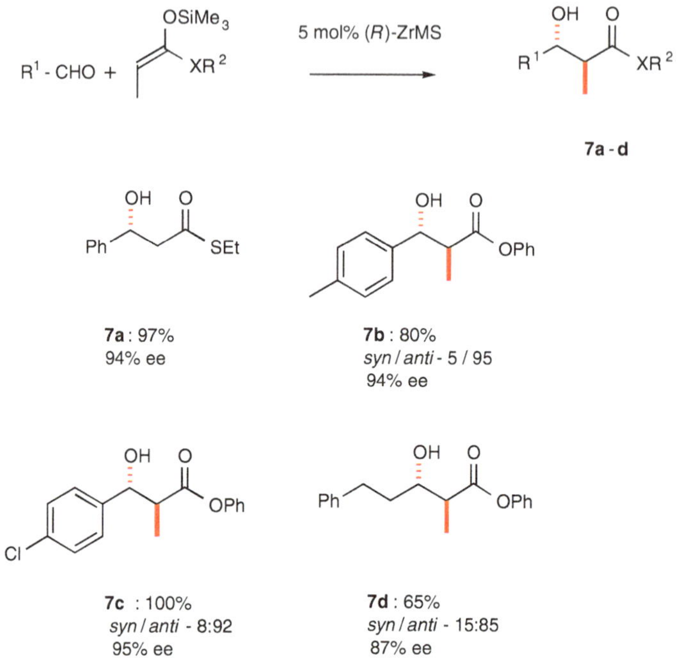

Scheme 3.1.6.5 Asymmetric aldol reactions in the presence of (*R*)–ZrMS complex

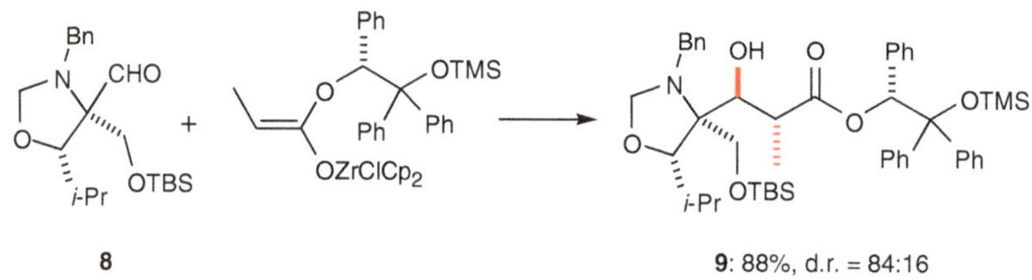

Scheme 3.1.6.6 Application of zirconium(IV) enolates in total synthesis of lactacystin

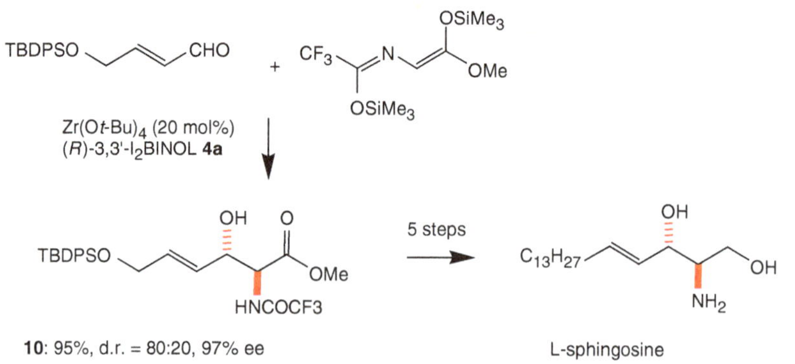

Scheme 3.1.6.7 Total synthesis of L-sphingosine in the presence of chiral zirconium catalyst

Scheme 3.1.6.8 Synthesis of a substituted aromatic trifluoroacetyl-serine methyl ester

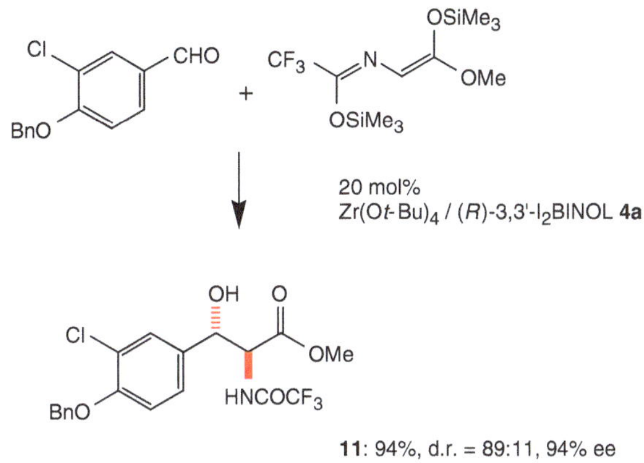

100 g-scale stereoselective production of (2*S*,3*S*)-p-benzyloxy-m-chlorophenyl-*N*-trifluoroacetyl-serine methyl ester **11** was achieved in high yields and high enantioselectivities when used with 5 mol% of air-stable chiral zirconium catalyst (*R*)–ZrMS (Scheme 3.1.6.8).[47]

References

1. Evans DA, McGee LR (1980) Tetrahedron Lett 21:3975
2. Yamamoto Y, Maruyama K (1989) Tetrahedron Lett 30:4607
3. Brown DW, Campbell MM, Taylor AP, Zhang XA (1987) Tetrahedron Lett 28:985
4. Panek JS, Bula OA (1988) Tetrahedron Lett 29:1661
5. Curran DP, Chao JC (1990) Tetrahedron 46:7325
6. Yamago S, Machii D, Nakamura E (1991) J Org Chem 56:2098
7. Wipf P, Xu W, Smitrovich JH (1994) Tetrahedron 50:1935
8. Evans DA, McGee LR (1981) J Am Chem Soc 103:2876
9. d'Angelo J, Pecquet-Dumas F (1983) Tetrahedron Lett 24:1403
10. Bernardi A, Colombo L, Gennari C, Prati L (1984) Tetrahedron 40:3769
11. Katsuki T, Yamaguchi M (1985) Tetrahedron Lett 26:5807
12. Sacha H, Waldmüller D, Braun M (1994) Chem Ber 127:1959
13. Vicario JL, Badia D, Dominguez E, Rodriguez M, Carrillo L (2000) J Org Chem 65:3754
14. Kurosu M, Lorca M (2001) J Org Chem 66:1205
15. Stork G, Shiner CS, Winkler JD (1982) J Am Chem Soc 104:310
16. Stork G, Winkler JD, Shiner CS (1982) J Am Chem Soc 104:3767
17. Sasai H, Kirio Y, Shibasaki M (1990) J Org Chem 55:5306
18. Hollis TK, Robinson NP, Bosnich B (1992) Tetrahedron Lett 33:6423
19. Hollis TK, Odenkirk W, Robinson NP, Whelan J, Bosnich B (1993) Tetrahedron 49:5415
20. Cozzi PG, Floriani C, Chiesi-Villa A, Rizzoli C (1994) Synlett 857
21. Cozzi PG, Floriani C (1995) J Chem Soc Perkin Trans 1:2557
22. Kanno K-I, Takahashi T In: Acid Catalysis in Modern Organic Synthesis, Yamamoto H, Ishihara K (eds). Wiley, Weinheim, vol 2, p 828
23. Ishitani H, Yamashita Y, Shimizu H, Kobayashi S (2000) J Am Chem Soc 122:5403
24. Yamashita Y, Ishitani H, Shimizu H, Kobayashi S (2002) J Am Chem Soc 124:3292
25. Parmee ER, Hong Y, Tempkin O, Masamune S (1992) Tetrahedron Lett 33:1729
26. Mikami K, Matsukawa S (1994) J Am Chem Soc 116:4077
27. Mahrwald R (1995) Tetrahedron 51:9015

28. Evans DA, MacMillan DWC, Campos KR (1997) J Am Chem Soc 119:10859
29. Yanagisawa A, Matsumoto Y, Nakashima H, Asakawa K, Yamamoto H (1997) J Am Chem Soc 119:9319
30. Denmark SE, Wong KT, Stavenger RA (1997) J Am Chem Soc 119:2333
31. Northrup AB, MacMillan DWC (2002) J Am Chem Soc 124:6798
32. Yanagisawa A, Matsumoto Y, Asakawa K, Yamamoto H (2002) Tetrahedron 58:8331
33. Denmark SE, Wynn T, Beutner GL (2002) J Am Chem Soc 124:13405
34. Wadamoto M, Ozawa N, Yanagisawa A, Yamamoto H (2003) J Org Chem 68:5593
35. Evans DA, Johnson DS (1999) Org Lett 1:595
36. Nakamura M, Hamashima Y, Usuda H, Kanai M, Shibasaki M (2000) Angew Chem Int Ed 39:1650
37. Evans DA, Scheidt KA, Johnston JN, Willis MC (2001) J Am Chem Soc 123:4480
38. Posner GH, Dai H, Bull DS, Lee JK, Eydoux F, Ishihara Y, Welsh W, Pryor N, Petr JrS (1996) J Org Chem 61:671
39. Terada M, Matsumoto Y, Nakamura Y, Mikami K (1997) Chem Commun 281
40. Terada M, Matsumoto Y, Nakamura Y, Mikami K (1999) Inorg Chim Acta 296:267
41. Ribe S, Wipf P (2001) Chem Commun 299
42. Kobayashi S, Saito S, Ueno M, Yamashita Y (2003) Chem Commun 2016
43. Sauvé G, Shwartz DA, Ruest L, Deslongchamps P (1984) Can J Chem 62:2929
44. Bernet B, Bishop PM, Caron M, Kawamata T, Roy BL, Ruest L, Sauvé G, Soucy P, Deslongchamps P (1985) Can J Chem 63:2810
45. Corey EJ, Reichardt GA, Kania R (1993) Tetrahedron Lett 44:6977
46. Kobayashi J, Nakamura M, Mori Y, Yamashita Y, Kobayashi S (2004) J Am Chem Soc 126:9192
47. Isoda T, Akiyama R, Oyamada H, Kobayashi S (2006) Adv Synth Catal 348:1813

3.1.7 Miscellaneous Lewis Acids

By application of copper or tin Lewis acids in Mukaiyama aldol reactions mostly *anti*-configured aldol adducts of enolsilanes, pyruvates and glyoxylates electrophiles were detected. In contrast to that scandium complexes proved to be effective catalysts in *syn*-selective aldol additions of enolsilanes and ethylglyoxylate (Scheme 3.1.7.1).[1]

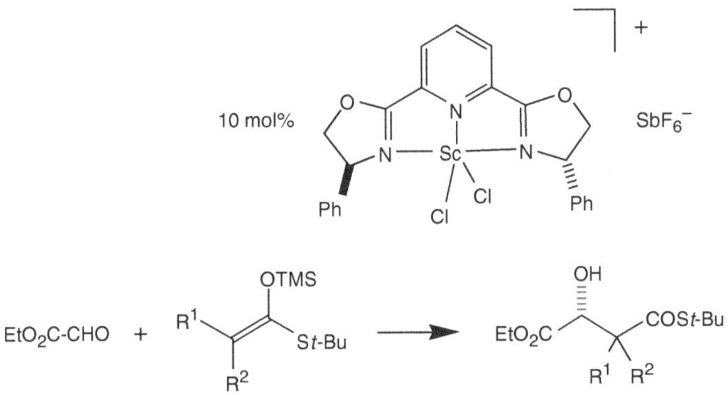

entry	R^1	R^2	yield (%)	*syn* / *anti*	ee (%)
1	Me	H	93	92 / 8	95
2	BnO	H	92	92 / 8	95
3	-$(CH_2)_3$-		90	-	95

Scheme 3.1.7.1 Scandium-catalysed glyoxylate aldol addition

A comparison between scandium- and copper-catalysed Mukaiyama aldol additions of ethylglyoxylate is given in Scheme 3.1.7.2. By deployment of copper catalyst in aldol additions using the same substrates the opposite enantiomer was detected with ee's over 98%.[2]

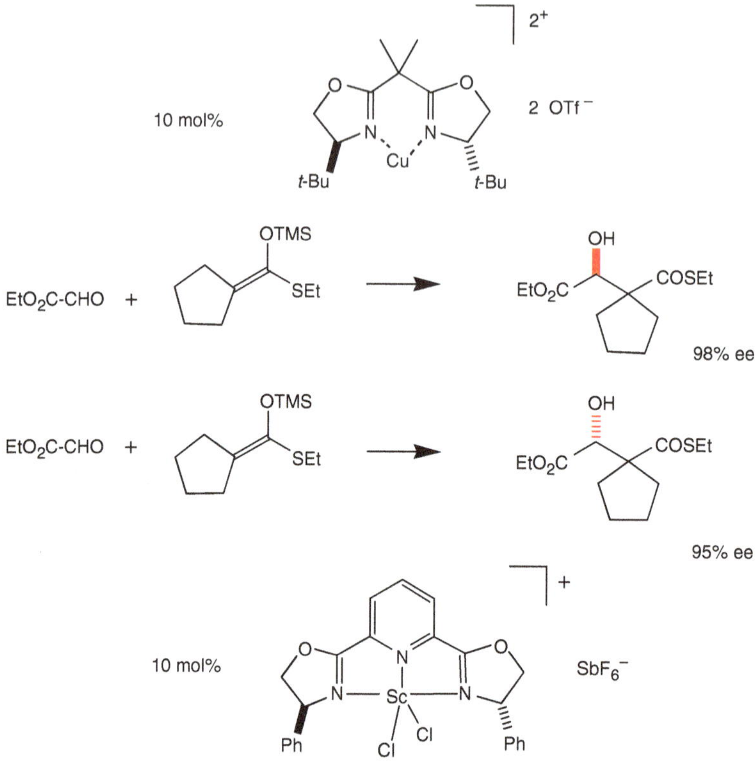

Scheme 3.1.7.2 Comparison between scandium- and copper-catalysed Mukaiyama aldol reactions of ethylglyoxylate

A detailed comparison between different lanthanide triflates in Mukaiyama aldol additions of pyruvate esters was published by Desimoni and coworkers. During these studies the scandium complexes provided highest yields. A correlation between enantioselectivities and ionic radius of lanthanides used in these reactions is also discussed (Scheme 3.1.7.3).[3]

An application of this method in total synthesis of optically active pantolactone derivatives was described by Evans and coworkers.[4] For an application of scandium(III) triflate in diastereoselective Mukaiyama reactions see Pohmakotr et al.[5]

Also, chiral nickel complexes are able to promote a direct aldol addition. In order to provide a catalytic execution this reaction was carried out in the presence of equimolar amounts of TMSOTf (Scheme 3.1.7.4).[6]

For applications of chiral chromium and aluminium salen complexes in vinylogous aldol additions see references.[7,8] For deployment of chiral iron- and zinc-Lewis acids in Mukaiyama reactions see reference.[9] For applications of chiral ytterbium complexes in Mukaiyama aldol reactions see Uotsu et al.[10] and for deployment of chiral lead complexes see reference.[11] Applications of indium-Lewis acids in Mukaiyama reactions were reported by Loh and coworkers.[12]

Scheme 3.1.7.3 Mukaiyama aldol reaction catalysed by chiral lanthanide complexes

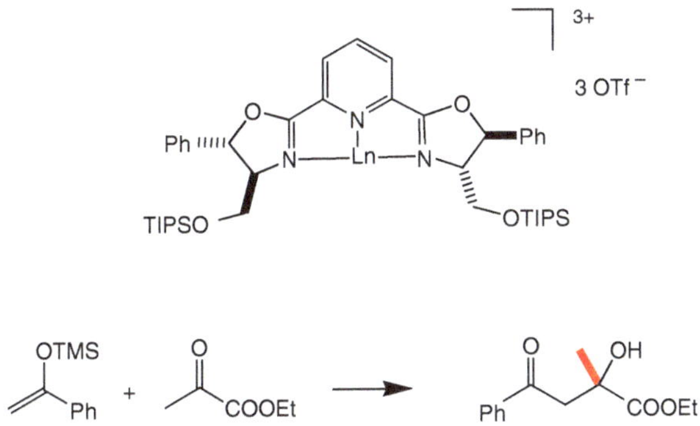

entry	Ln	ionic radius	yield (%)	ee (%, conf.)
1	Sc	0.870	92	92 (S)
2	Lu	0.977	70	95 (S)
3	Yb	0.985	55	79 (S)
4	Ho	1.015	45	24 (S)
5	Y	1.019	41	48 (S)
6	Eu	1.066	39	39 (S)
7	La	1.160	45	72 (R)

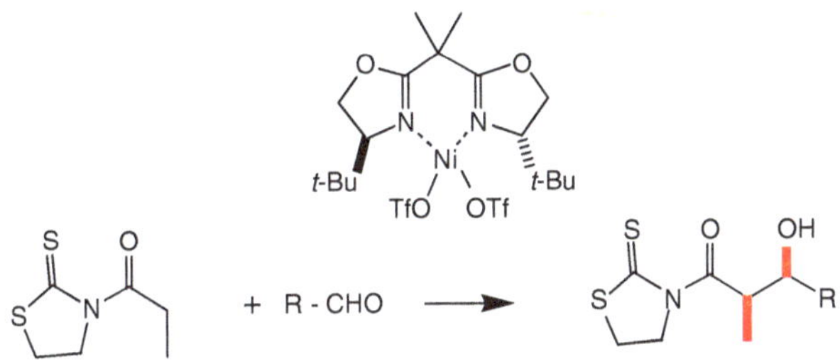

entry	R	*syn*/*anti*	yield (%)	ee (%)
1	Ph	94 / 6	81	97
2	4-Me-C_6H_4	93 / 7	80	95
3	Ph-CH = CH	88 /12	63	93
4	Me	97 / 3	86	93
5	*i*-Bu	98 / 2	70	90
6	*n*-Pr	97 / 3	84	90

Scheme 3.1.7.4 Enantioselective nickel-catalysed aldol addition to various aldehydes. 10 mol% nickel complex, 3.3 equiv. lutidine, 1.3 equiv. TMSOTf

References

1. Evans DA, Masse CE, Wu J (2002) Org Lett 4:3375
2. Evans DA, Kozlowski MC, Murry JA, Burgey CS, Campos KR, Connell BT, Staples RJ (1999) J Am Chem Soc 121:669
3. Desimoni G, Faita G, Piccinini F, Toscanini M (2006) Eur J Org Chem 5528
4. Evans DA, Wu J, Masse CE, MacMillan DWC (2002) Org Lett 4:3379
5. Pohmakotr M, Yotapan N, Tuchinda P, Kuharkan C, Reutrakul V (2007) J Org Chem 72:5016
6. Evans DA, Downey CW, Hubbs JL (2003) J Am Chem Soc 125:8706
7. (a) Shimada Y, Matsuoka Y, Irie R, Katsuki T (2004) Synlett 57; (b) Onitsuka S, Matsuoka Y, Irie R, Katsuki T (2003) Chem Lett 32:974; (c) Matsuoka Y, Irie R, Katsuki T (2003) Chem Lett 32:584
8. Evans DA, Janey JM, Magomedov N, Tedrow JS (2001) Angew Chem Int Ed 40:1884
9. (a) Jankowska J, Paradowska J, Mlynarski J (2006) Tetrahedron Lett 47:5281; (b) Jankowska J, Mlynarski J (2006) J Org Chem 71:1317
10. Uotsu K, Sasai H, Shibasaki M (1995) Tetrahedron: Asymm 6:71
11. (a) Manabe K, Kobayashi S (2002) Chem Eur J 8:4095; (b) Nagayama S, Kobayashi S (2000) J Am Chem Soc 122:11531
12. Fu F, Teo YC, Loh TP (2006) Tetrahedron Lett 47:4267

3.2 Lewis Base-Catalysed Aldol Additions

In 1996, Denmark et al. described an aldol reaction of trichlorosilyl enolates with aldehydes for the first time.[1] In contrast to the well-established Mukaiyama reaction, where catalytic amounts of Lewis acids were deployed in reactions with trimethylsilyl enolates, this transformation was catalysed by Lewis bases. When used with chiral Lewis bases, aldol adducts can be obtained with a high degree of enantioselectivity. For several comprehensive overviews in this field see Denmark and coworkers.[2,3,4] For a general and comprehensive overview of Lewis base catalysis in organic chemistry see Denmark and Beutner.[5]

The development of Lewis base-catalysed aldol additions was strongly connected with the development of an easy and general protocol for the synthesis of trichlorosilyl enolates and chiral Lewis bases. These problems were solved by Denmark and coworkers. They developed several useful and general procedures for the preparation of trichlorosilyl enolates of carbonyl compounds by metal exchange reactions. These are reactions between corresponding enol stannanes of ketones with $SiCl_4$, or metal exchange of trimethylsilyl enolethers with $SiCl_4$ in the presence of catalytic amounts of mercury(II) or tin(IV) salts or $Pd(OAc)_2$.[6] The synthesis of trichlorosilyl enolates of aldehydes was accomplished by transmetallation of the corresponding trimethylsilyl ethers of aldehydes with $SiCl_4$ in the presence of catalytic amounts of $Pd(OAc)_2$.[7] Trichlorosilyl enolates of ketones react at room temperature with aldehydes without any activation (Scheme 3.2.1).[8]

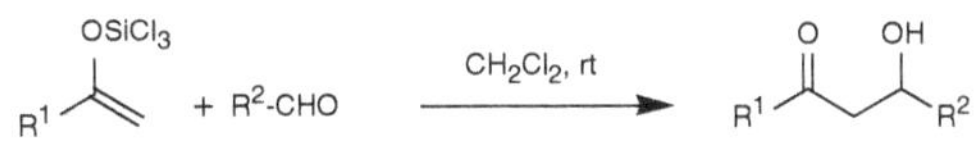

1a-e

Scheme 3.2.1 Noncatalysed aldol reaction of trichlorosilyl enolates at room temperature

compound	R^1	R^2	yield (%)
1a	Me	Ph	92
1b	*n*-Bu	Ph	95
1c	*n*-Bu	Cy	93
1d	Ph	Ph	91
1e	*i*-Pr	Ph	93

High yields of the corresponding aldol adducts were obtained after 4–10h at room temperature. In reactions of pivaldehyde with the trichlorosilyl enol ether of methyl butyl ketone only traces of aldol products were detected. By addition of catalytic amounts of HMPA nearly quantitative transformations could be obtained.[9] Subsequent optimization of this observation led to the following general protocol. The reactions were carried out at −78°C. Without any additives only traces of aldol adducts were detected, whereas in the presence of several chiral phosphoramides good yields to full conversations were obtained. Results of aldol additions with benzaldehyde in the presence of 10mol% of different chiral phosphoramides **A–D** are shown in Scheme 3.2.2.

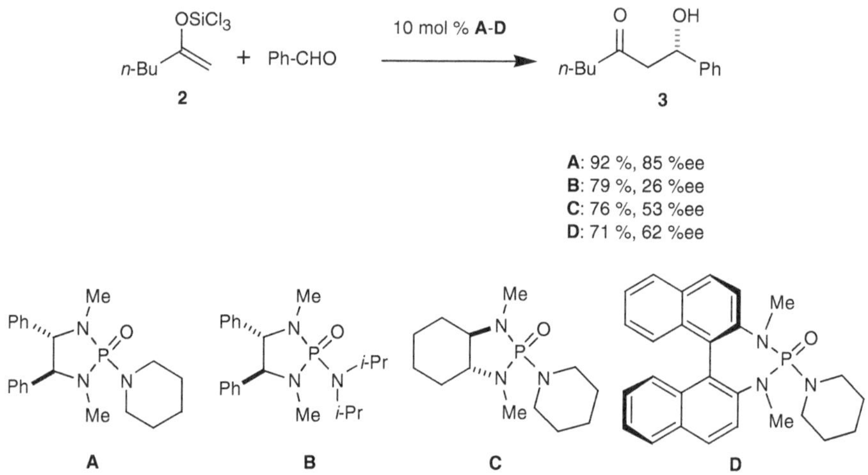

Scheme 3.2.2 Chiral phosphoramides as catalysts in enantioselective aldol additions of trichlorosilyl enol ethers to benzaldehyde

When used with 5mol% of the chiral diamine-derived phosphoramide **A** the acetate aldol adducts of a variety of trichlorosilyl enolates with benzaldehyde were isolated in high yields and good enantioselectivities (Scheme 3.2.3).[10]

Also, cyclic trichlorosilyl enol ethers derived from cyclohexanone react with several aldehydes in the presence of catalytic amounts of chiral phosphoramide to give the expected aldol adducts in high yields – nearly quantitative – and good to excellent enantioselectivities.[11] Unfortunately, reactions of enolizable aldehydes do not result in the formation of the corresponding aldol products. Results of reactions in the presence of phosphoramide A are shown in Scheme 3.2.4.

A correlation can be observed between the geometry of the starting enol ethers and the diastereoselection detected in the aldol adduct. Similar to the behaviour of boron enolates in aldol additions (*Z*)-trichloro enolates of diethylketone gives *syn*-configured aldol adducts selectively, whereas the use of (*E*)-trichlorsilyl enolates provides the *anti*-configured aldol adducts (Scheme 3.2.5).[12]

Scheme 3.2.3 Stereochemical results in acetate aldol additions with benzaldehyde

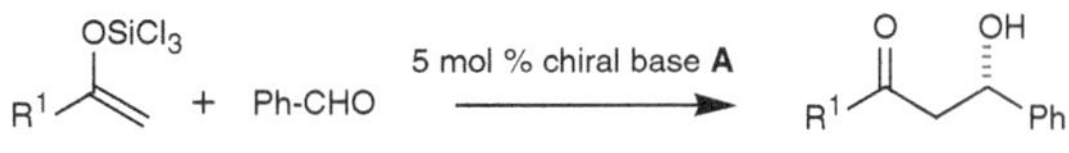

4a-g

compound	R^1	yield (%)	ee (%)
4a	Me	98	87
4b	*n*-Bu	98	33
4c	*i*-Bu	95	82
4d	*i*-Pr	97	81
4e	*t*-Bu	95	52
4f	Ph	93	49
4g	TBSO	94	86

Scheme 3.2.4 Lewis base-catalysed aldol addition of trichlorosilyl enol ethers of cyclohexanone to aromatic aldehydes

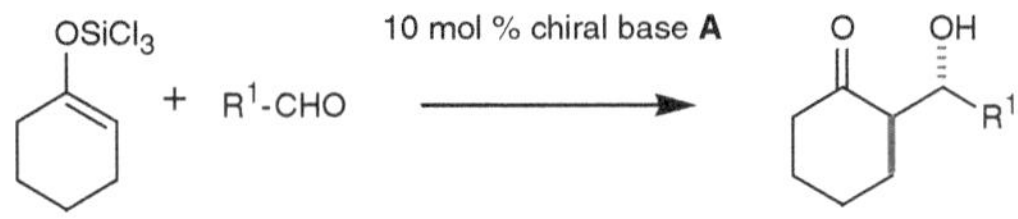

5a-e

compound	R^1	yield (%)	*syn* / *anti*	ee (%)
5a	Ph	95	2 / 98	93
5b	PhCH = CH	94	1 / 99	88
5c	PhC ≡ C	90	16 / 84	82
5d	1-napthyl	94	1 / 99	97
5e	PhCH = CMe	98	1 / 99	92

Scheme 3.2.5 Aldol additions of (*Z*)- or (*E*)-trichlorosilyl enol ethers of diethylketone

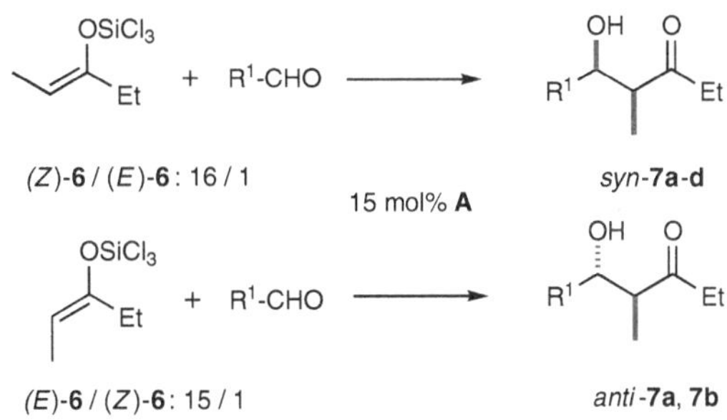

compound	R^1	yield (%)	*syn* / *anti*	ee (%)
7a	Ph	84	94 / 6	91
7b	2-furyl	79	80 / 20	91
7c	PhCH = CH	85	83 / 17	91
7d	Cy	45	94 / 7	33
anti-**7a**	Ph	86	50 / 50	78
anti-**7b**	2-furyl	76	33 / 67	50

This transformation can be extended to enantioselective cross-aldol additions of trichlorosilyl enol ethers of aldehydes to other aldehydes.[13] Reactions were carried out at −65°C in the presence of chiral phosphoramides. In reactions with benzaldehyde high stereoselectivities were obtained. Again, the optional installation of *syn-* or *anti*-diastereoselectivity is given by the deployment of (*Z*)- or (*E*)-trichlorosilyl enol ethers of the starting aldehydes. The optical active β-hydroxyaldehydes were isolated as their corresponding acetals **9** (Scheme 3.2.6).

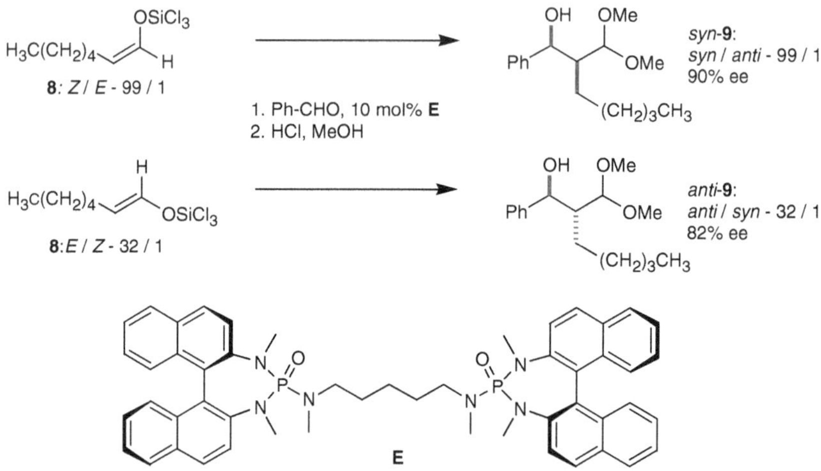

Scheme 3.2.6 Aldol additions of (*Z*)- or (*E*)-trichlorosilyl enol ethers of enolizable aldehydes

For a Lewis base-catalysed cross-aldol addition of trimethylsilyl enol ether of acetaldehyde to aromatic aldehydes see Denmark and Bui.[14] Even enolizable aldehydes can be used in these reactions. Slow additions and longer reaction times are essential for good yields and selectivities (syring pump technique). Excellent diastereoselectivities and moderate to good enantioselectivities were detected under these conditions.[15] The intermediary aldehydes were isolated as their acetals. *Syn*-configured acetals **11a–f** were observed using the corresponding (*Z*)-configured trichlorosilyl enol ethers of aldehydes (Scheme 3.2.7).

In contrast to these results high *anti*-diastereoselectivities were observed when used with the corresponding (*E*)-configured trichlorosilyl enol ethers (Scheme 3.2.8).

For Lewis base-catalysed aldol additions of aldehydes with trichlorosilyl enol ethers of α-oxy-aldehydes see Denmark and Ghosh.[16]

Also, reactions of trichlorosilyl enol ethers of acetic acid esters with ketones were carried out. These enantioselective reactions were mediated by the use of chiral pyridine-*N*-oxides. Thus, an approach to optically active tertiary alcohols is given. The quaternary stereogenic centre was constructed with moderate to good enantioselectivities depending on the ketone used (Scheme 3.2.9).[17]

Scheme 3.2.7 *syn*-Selective cross-aldol additions of aldehydes with (*Z*)-configured trichlorosilyl enol ether of propionaldehyde

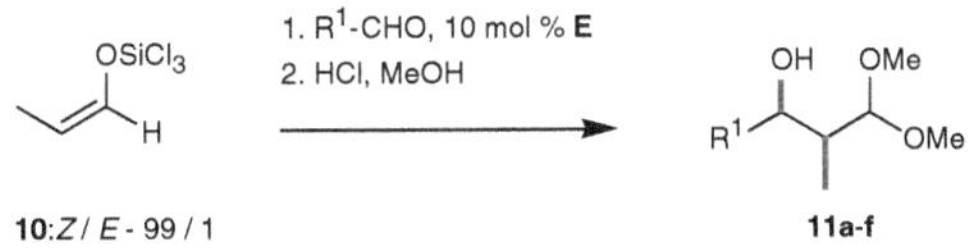

compound	R^1	yield (%)	*syn* / *anti*	ee (%)
11a	Ph	95	98 / 2	81
11b	PhCH = CH	86	99 / 1	41
11c	PhCH = CMe	91	97 / 3	68
11d	PhC ≡ C	98	98 / 2	9
11e	$PhCH_2CH_2$	47	95 / 5	9
11f	Cy	42	97 / 3	44

Scheme 3.2.8 *anti*-Selective cross-aldol additions of aldehydes with (*E*)-configured trichlorosilyl enol ether of propionaldehyde

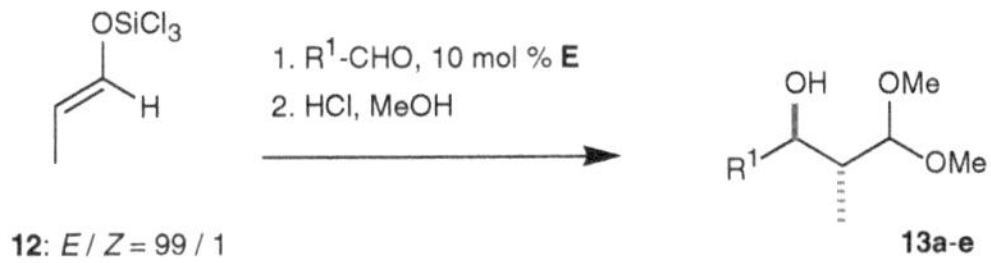

compound	R^1	yield (%)	*syn* / *anti*	ee(%)
13a	Ph	97	1 / 99	59
13b	PhCH = CH	88	1 / 99	26
13c	PhC ≡ C	99	2 / 98	76
13d	$PhCH_2CH_2$	79	1 / 99	80
13e	Cy	69	1 / 99	67

Scheme 3.2.9 Enantioselective aldol additions of trichlorosilyl enol ethers of acetic acid methyl ester to ketones

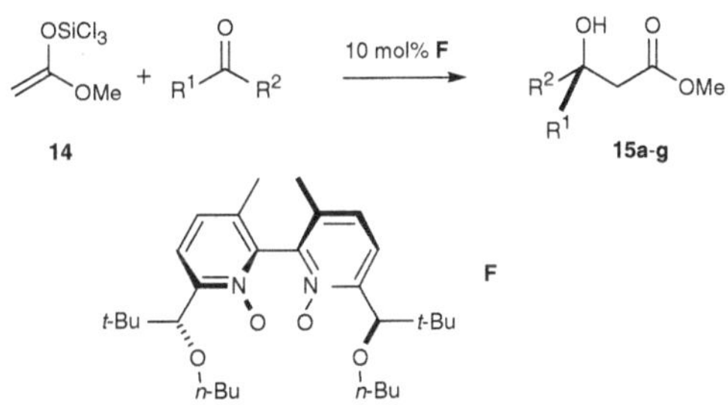

compound	R^1	R^2	yield (%)	ee(%)
15a	Ph	Me	96	82
15b	Ph	C ≡ C	89	86
15c	naphtyl	Me	89	56
15d	2-furyl	Me	87	49
15e	$PhCH_2CH_2$	Me	97	35
15f	Cy	Me	91	31
15g	*t*-Bu	Me	97	43

An in situ performance of these transformations discussed above was given by the application of trialkylsilyl enol ethers in aldol reactions with equimolar amounts of $SiCl_4$ and catalytic amounts of chiral phosphoramides. Thus, aldol adducts were obtained in nearly quantitative yields and good to excellent enantioselectivities (Scheme 3.2.10).

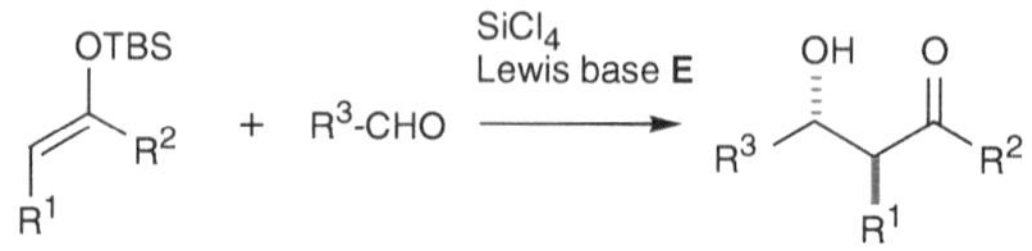

16a-h

compound	R^1	R^2	R^3	yield (%)	*syn* / *anti*	ee (%)
16a	H	OMe	Ph	97	-	93[a]
16b	H	OMe	2-furyl	94	-	87[a]
16c	Me	OMe	Ph	98	1 / 99	72[b]
16d	Me	O*t*-Bu	Ph	93	1 / 99	98[b]
16e	Me	OEt	$Ph(CH_2)_2$	71	9 / 91	88[c]
16f	Me	OEt	Cy	49	11 / 89	35[d]
16g	H	*n*-Bu	Ph	99	-	>98[e]
16h	H	*n*-Bu	2-furyl	88	-	90[e]

Scheme 3.2.10 Lewis base-catalysed enantioselective aldol addition of various silyl enol ethers to different aldehydes reaction conditions: [a]5 mol% **E**, 110 mol% $SiCl_4$; [b]1 mol% **E**, 110 mol% $SiCl_4$; [c]5 mol% **E**, 110 mol% $SiCl_4$, 10 mol TBAI; [d]10 mol% **E**, 110 mol% $SiCl_4$, 10 mol TBAI; [e]5 mol% **E**, 150 mol% $SiCl_4$, iPr_2NEt

Tert-butyldimethylsilyl enol ethers of ketones,[18] carboxylic esters,[19,20] amides[21] and vinylogous esters[22,23,24,25] reacted with a variety of aldehydes under these conditions.

It is worthy to note that the geometry of ketene acetals used in these reactions does not play any role to the diastereoselectivity observed in these aldol adducts. This is in strong contrast to what was observed in reactions discussed in Schemes 3.2.7 and 3.2.8. Independent of the geometry of starting ketene acetals the *anti*-configured aldol adducts were isolated with an excellent degree of diastereoselectivity. Enolizable aldehydes do not react with ketene acetals or silyl enol ethers of ketones under these conditions.

Also, trichlorosilyl enol ethers of α- and β-chiral ketones were deployed to several Lewis base-catalysed aldol additions. Noncatalysed additions to α-chiral ketones were unsuccessful with regard to stereoselectivities. Only modest substrate-induced stereoselectivities were measured. When used with chiral phosphoramides in acetate aldol additions an extremely matched/mismatched situation occurred.[26] Depending on the chirality of the catalyst used, high diastereoselectivities were obtained. Moreover, the degree of diastereoselectivity obtained depends on the nature of α-substituents as well.[27] Results of application of chiral phosphoramide **A** are shown in Scheme 3.2.11. For results in the 'propionate' aldol series see also references.[28,29]

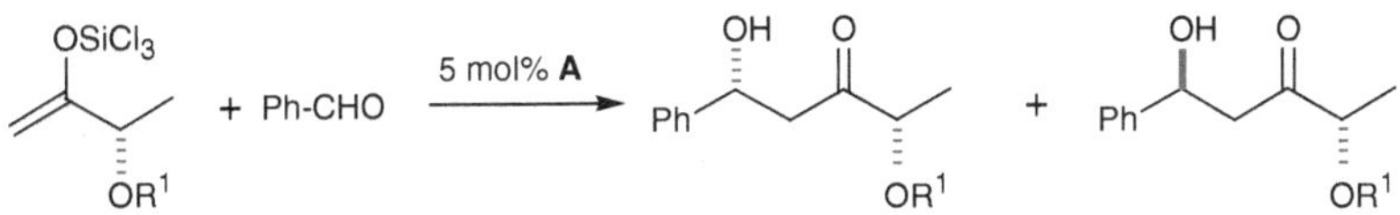

syn-**17a-f** anti-**17a-f**

compound	R^1	yield (%)	syn / anti
17a	TBS	85	99 / 1[a]
17b	TBS	85	60 / 40[b]
17c	Piv	78	95 / 5[a]
17d	Piv	78	77 / 23[b]
17e	Bn	77	52 / 48[a]
17f	Bn	78	48 / 52[b]

Scheme 3.2.11 Matched and mismatched situations in reactions of trichlorosilyl enol ethers of chiral ketones with benzaldehyde reaction conditions: [a]5 mol% (*R*,*R*)-**A**; [b]5 mol% (*S*,*S*)-**A**

Even trichlorosilyl enol ethers of β-chiral ethyl ketones react with both enantiomers of chiral catalysts to give the corresponding diastereoisomers with high induced or internal stereoselection (Scheme 3.2.12).[30,31]

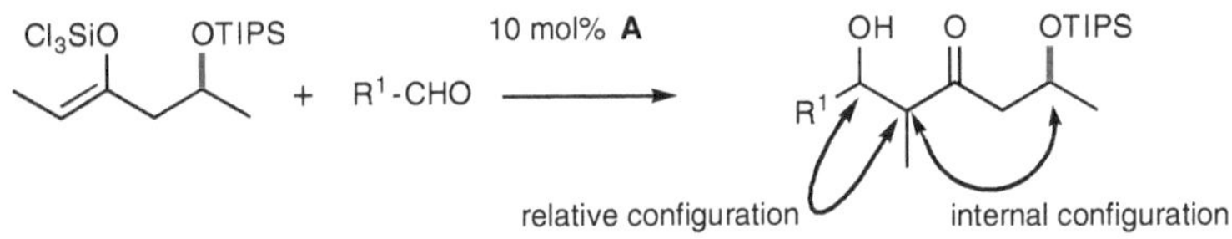

18 **19a-f**

compound	R^1	yield (%)	*syn* / *anti* (internal)	*syn* / *anti* (relative)
19a	Ph	84	6 / 94	97 / 3[a]
19b	Ph	86	91 / 9	96 / 4[b]
19c	MeCH = CH	79	12 / 88	97 / 3[a]
19d	MeCH = CH	83	14 / 86	97 / 3[b]
19e	PhCH = CH	80	9 / 91	98 / 2[a]
19f	PhCH = CH	75	88 / 12	98 / 2[b]

Scheme 3.2.12 Internal and relative stereoselectivity in the propionate aldol series reaction conditions: [a]10 mol% (*R*,*R*)-**A**; [b]10 mol% (*S*,*S*)-**A**

Also, optically active 1,2-diols are accessible by this method. By an enantioselective glycolate aldol addition with aldehydes in the presence of chiral phopshoramide **E** the aldol adducts were isolated with high enantioselectivities. These products represent valuable precursors for optically active 1,2-diols.[32] Recently, BINAPO-catalysed aldol additions of trichlorosilyl enol ethers were described. The *anti*-configured aldol products were obtained with high degrees of enantioselectivity.[33]

After these systematic and fundamental works of Denmark and coworkers a series of reports were published to demonstrate the utility of this method. For example, Denmark et al. used their own elaborated method to synthesize macrolide RK-397. The starting chiral aldol adduct **22** was obtained by a vinylogous aldol reaction of ketene acetal **20** and α, β-unsaturated aldehyde **21** using chiral phosphoramide **E** in the presence of $SiCl_4$ (96% ee) (Scheme 3.2.13). The second aldol step of methyl ketone **23** and α,β-unsaturated aldehyde **24** was achieved by employment of chiral phosphoramide **A** and $SiCl_4$. For comprehensive reviews of applications of vinylogous aldol additions see Denmark et al. and others.[34]

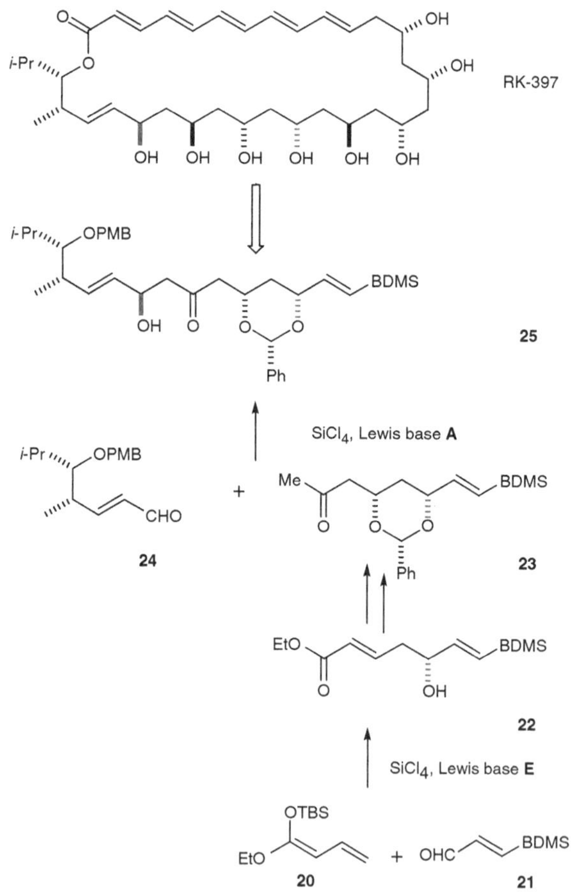

Scheme 3.2.13 Total synthesis of RK-397

Notably, the diastereoselectivity was only 67/33 during the vinylogous aldol reaction favouring the required product **25**. Better diastereoselectivities were obtained when used the established boron enolate methodology (dr > 95/5).[35]

Also, Lewis bases were reported as catalysts in Mukaiyama reactions (Scheme 3.2.14).[36,37,38] The authors used catalytic amounts of lithiated pyrrolidones in aldol additions of silyl enol ethers to aromatic aldehydes.

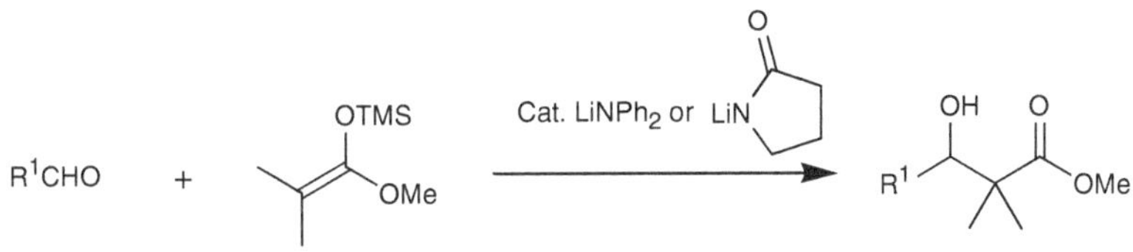

compound	R^1	yield (%)
26a	4-MeC_6H_4	95
26b	4-MeO-C_6H_4	95
26c	4-CN-C_6H_4	89
26d	4-Cl-C_6H_4	95
26e	4-Me-C_6H_4	95
26f	4-Me-C_6H_4	95

Scheme 3.2.14 Application of lithiated amides in Mukaiyama aldol additions

Also, trialkyloxysilyl enol ethers were used in enantioselective Lewis base-catalysed Mukaiyama reactions. Trimethoxysilyl enol ethers are known to be more stable than the corresponding trichlorosilyl enol ethers. In the presence of catalytic amounts of dilithium salt of chiral BINOL the corresponding aldol adducts were observed with a high degree of enantioselectivity.[39]

Quaternary ammonium salts[40] and lithium salts[41] were described as Lewis base catalysts in Mukaiyama reactions. Furthermore, amine *N*-oxides,[42] *N*-methylimidazole[43] and sodium phenoxide/phosphine oxides[44] were reported as Lewis base catalysts in Mukaiyama reaction.

References

1. Denmark SE, Winter SBD, Su X, Wong KT (1996) J Am Chem Soc 118:7404
2. Denmark SE, Stavenger RA (2000) Acc Chem Res 33:432
3. Denmark SE, Bui T (2004) Proc Natl Acad Sci USA 101:5439
4. Denmark SE, Fujimori S (2004) In: Modern Aldol Reactions, Mahrwald R (ed). Wiley-VCH, Weinheim, vol 2, p 229

5. Denmark SE, Beutner GL (2008) Angew Chem Int Ed 47:1560
6. Denmark SE, Stavenger RA, Winter SBD, Wong KT, Barsanti P (1998) J Org Chem 63:9517
7. Denmark SE, Ghosh SK (2001) Angew Chem Int Ed 40:4759
8. Denmark SE, Stavenger RA, Wong KT, Su X (1999) J Am Chem Soc 121:4982
9. Denmark SE, Stavenger RA, Wong KT (1998) J Org Chem 63:918
10. Denmark SE, Stavenger RA (2000) J Am Chem Soc 122:8837
11. Denmark SE, Stavenger RA, Wong KT (1998) Tetrahedron 54:10389
12. Denmark SE., Pham SM (2003) J Org Chem 68:5045
13. Denmark SE, Ghosh SK (2001) Angew Chem Int Ed 40:4759
14. Denmark SE, Bui T (2005) J Org Chem 70:10190
15. Denmark SE, Bui T (2004) Proc Natl Acad Sci USA 101:5439
16. Denmark SE, Ghosh SK (2007) Tetrahedron 63:8636
17. Denmark SE, Fan Y (2002) J Am Chem Soc 124:4233
18. Denmark SE, Heemstra JR Jr (2003) Org Lett 5:2303
19. Denmark SE, Wynn T, Beutner GL (2002) J Am Chem Soc 124:13405
20. Denmark SE, Fan Y, Eastgate MD (2005) J Org Chem 70:5235
21. Denmark SE, Heemstra JR (2006) J Am Chem Soc 128:1038
22. Denmark SE, Beutner GL (2003) J Am Chem Soc 125:7800
23. Villano R, Acocella MR, Massa A, Palombi L, Scettri A (2006) Tetrahedron: Asymm 17:3332
24. Palombi L, Acocella MR, Celenta N, Massa A, Villano R, Scettri A (2006) Tetrahedron Asymm 17:3330
25. Denmark SE, Heemstra JR Jr (2007) J Org Chem 72:5668
26. Denmark SE, Stavenger RA (1998) J Org Chem 63:9524
27. Denmark SE, Fujimori S (2001) Synlett 1024
28. Denmark SE, Fujimori S (2002) Org Lett 4:3473
29. Denmark SE, Pham SM (2001) Org Lett 3:2201
30. Denmark SE, Fujimori S (2002) Org Lett 4:3477
31. Denmark SE, Fujimori S, Pham SM (2005) J Org Chem 70:10823
32. Denmark, SE, Chung WJ (2008) Angew Chem Int Ed 47:1890
33. Kotani S, Hashimoto S, Nakajima M (2006) Synlett 1116
34. (a) Denmark SE, Heemstra JR Jr, Beutner GL (2005) Angew Chem Int Ed 44:4682; (b) Kalesse M (2005) In: Topics in Current Chemistry, Mulzer JH (ed). Springer-Verlag, Berlin/Heidelberg, 244:43; (c) Hosokawa S, Tatsuta K (2008) Mini-Rev Org Chem 5:1
35. Denmark SE, Fujimori S (2005) J Am Chem Soc 127:8971
36. Fujisawa H, Mukaiyama T (2002) Chem Lett 182
37. Fujisawa H, Mukaiyama T (2002) Chem Lett 858
38. Mukaiyama T, Fujisawa H, Nakagawa T (2002) Helv Chim Acta 85:4518
39. (a) Nakajima M, Orito Y, Ishizuka T, Hashimoto S (2004) Org Lett 6:3763; (b) Orito Y, Hashimoto S, Ishizuka T, Nakajima M (2006) Tetrahedron 62:390; (c) Ichibakase T, Orito Y, Nakajima M (2008) Tetrahedron Lett 49:4427
40. (a) Mizugaki T, Hetrick CE, Murata M, Ebitani K, Amiridis M, Kaneda K (2005) Chem Lett 34:420; (b) Fujisawa H, Nagata Y, Sato Y, Mukaiyama T (2005) Chem Lett 34:842
41. Kawano Y, Fujisawa H, Mukaiyama T (2005) Chem Lett 34:614
42. Hagiwara H, Inoguchi H, Fukushima M, Hoshi T, Suzuki T (2005) Synlett 2388
43. Hagiwara H, Inoguchi H, Fukushima M, Hoshi T, Suzuki T (2006) Tetrahedron Lett 47:5371
44. Hatano M, Takagi E, Ishihara K (2007) Org Lett 9:4527

3.3 Direct Aldol Addition

Several aspects of C–C bond formation processes in the mid-1990s led to the beginning of the development of so-called direct aldol additions. Due to the increasing demands of the industry for environmentally clean and economic reaction conditions and atom economy[1] stoichiometric amounts of reagents should be excluded. These processes are connected with waste, salts, aqueous solutions of base or acids. For that reasons a direct and catalytic asymmetric aldol addition is strongly suggested.

Early examples of this development are the applications of catalytic amounts of Lewis acids using unactivated carbonyl components in direct aldol additions. Miyoshi and coworkers described the employment of a $BiCl_3$-Zn reagent.[2] Mahrwald and coworkers reported the use of substoichiometric amounts of titanium(IV) halides in direct aldol additions.[3,4] When used with unsymmetrical ketones in these direct aldol additions high degrees of regio- and diastereoselectivities were measured. C–C bond formation was observed at the sterically more encumbered α-position of the ketones only.[5] Later on Tanabe and coworkers reported the in situ generation of silyl enol ethers using catalytic amounts of trimethylchlorosilane in the presence of a $TiCl_4$/amine reagent. Thus, the authors were able to react ketones with sterically overcrowded ketones.[6,7,8,9]

The prototype of an enantioselective and catalytic performance of an aldol addition is given by Nature. Aldolases catalyse the direct catalytic aldol addition under very mild conditions.[10] These processes are found in several carbohydrate metabolisms. With regard to the reaction mechanism these enzymes are divided in two classes. Class I aldolases are found mainly in plants and mammalians. The reaction mechanism based on an amine catalysis. The ene component is activated by a lysine residue of the active site of the enzyme forming an imine-enamine. A stereospecific reaction with an aldehyde yields after hydrolysis the aldol adduct. Class II aldolases utilize zinc ions as Lewis acids to activate their substrates. At the same time a tyrosine residue from the adjoining subunit simultaneously assists in the activation of the incoming aldehyde. Thus, nature achieves a highly catalytic as well as enantioselective aldol reaction without any need for separate reaction steps to form the enolate or the enol ether. These enzymes are found in fungi and bacteria.

These working models of aldolases have been inspiring chemists for a long time. The mode of action of class I aldolases acts as a model for all kinds of

R. Mahrwald, *Aldol Reactions*,

organocatalytic aldol processes,[11,12,13] whereas the reaction mechanism of class II aldoldases serves as a model for achieving the direct, catalytic and enantioselective aldol additions.[14] In 1997, Shibasaki and coworkers were the first to succeed in realizing this concept.[15] They reported the synthesis of a heterobimetallic lanthanum–lithium–BINOL complex (LLB) and could demonstrate its application in the direct catalytic asymmetric aldol addition. Several aliphatic aldehydes were reacted with methylaryl ketones and methylalkyl ketones. Generally, a great excess of ketone is necessary to achieve the desired conversion, and complete conversion requires several days, sometimes a week or even longer (Scheme 3.3.1).

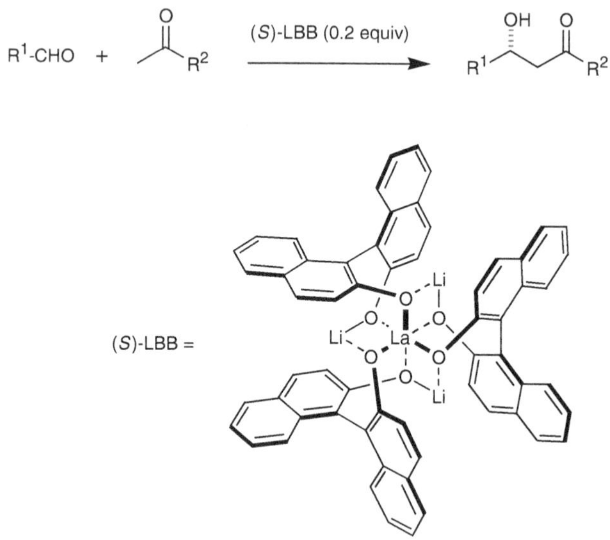

entry	R^1	R^2	equiv. ketone	time (h)	yield (%)	ee (%)
1	*t*-Bu	Ph	5	88	76	88
2	*t*-Bu	Ph	1.5	135	43	87
3	*t*-Bu	Ph	10	91	81	91
4	*t*-Bu	1-naphthyl	8	253	55	76
5	*t*-Bu	Me	10	100	53	73
6	$PhCH_2C(CH_3)_2$	Ph	7.4	87	90	69
7	$PhCH_2C(CH_3)_2$	Me	10	185	82	74[a]
8	$PhCH_2C(CH_3)_2$	Et	50	185	71	94
9	Cy	Ph	8	169	72	44
10	*i*-Pr	Ph	8	277	59	54
11	$PhCH_2CH_2$	Ph	10	72	28	52

Scheme 3.3.1 Direct asymmetric aldol reaction of methyl ketones using a heterobimetallic lanthanum–lithium–(*S*)–BINOL complex. [a]The reaction was carried out at –30°C

The (*S*)–LLB complex was employed to the total synthesis of fostriecin (Scheme 3.3.2).[16,17] By using 10 mol% of the bimetallic complex (*S*)–LLB the authors were able to isolate the acetylenic ketone **3** in good yields.

Scheme 3.3.2 Total synthesis of fostriecin

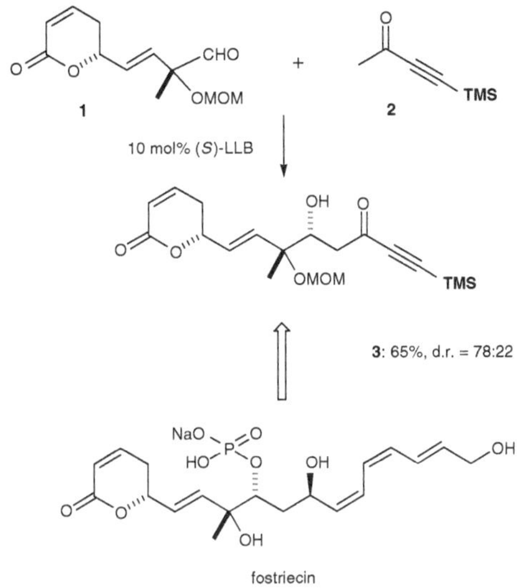

A further application of this methodology was found in the synthesis of key intermediate **6** of epothilone. The required configuration at C3 could be installed by a (*R*)-LLB-catalysed direct aldol addition of racemic aldehyde **4** and acetophenone (Scheme 3.3.3).[18] Aldol adduct **6** was isolated together with its diastereoisomer **7** via a catalytic resolution of the starting racemic aldehyde **4**.

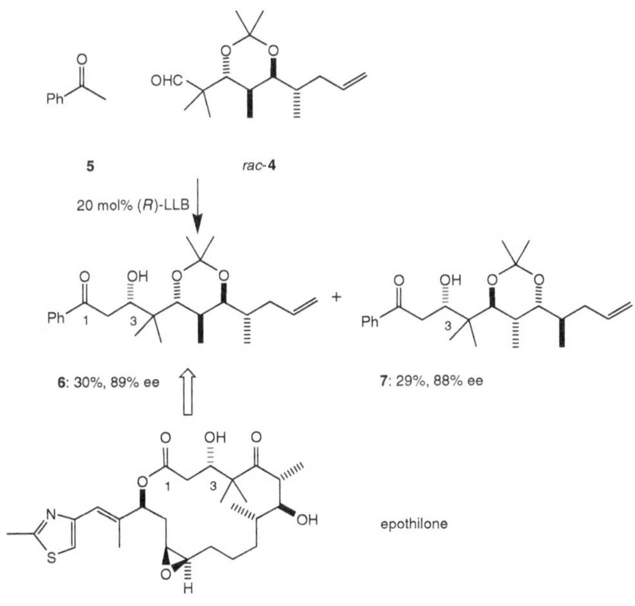

Scheme 3.3.3 Total synthesis of epothilone

An improvement of the LBB-methodology was achieved by the use of the LLB–KOH complex. This complex was prepared from LLB complex, KHMDS and H_2O. When used with this complex shorter reaction times and diminishing of catalyst loading in direct aldol additions are possible.[19] By using aromatic hydroxyketones as ene components the corresponding 1,2-diols were obtained in high yields and enantioselectivities. Moreover, by using 5–10 mol% of (*S*)–LLB–KOH complex an *anti*-preference is observed (Scheme 3.3.4).[20]

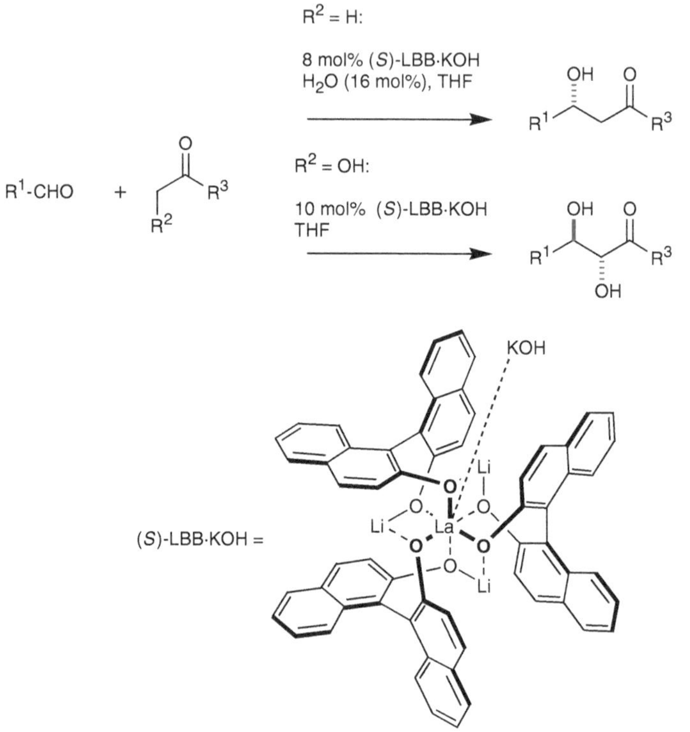

entry	R^1	R^2	R^3	yield (%)	*anti* / *syn*	ee (%) *anti* / *syn*
1	*t*-Bu	H	Ph	85	-	89
2	*t*-Bu	H	Me	62	-	76
3	*t*-Bu	H	Et	72	-	88
4	$PhCH_2C(CH_3)_2$	H	Ph	83	-	85
5	$BnOCH_2C(CH_3)_2$	H	Ph	91	-	90
6	*i*-Pr	H	Ph	90	-	33
7	$Ph(CH_2)_3$	OH	Ph	84	84 / 16	95 / 74
8	*i*-Bu	OH	Ph	86	65 / 35	90 / 83

Scheme 3.3.4 *anti*-Selective direct asymmetric aldol reaction catalysed by a lanthanum–lithium–(S)–BINOL–KOH–complex (LBB)

Later on, Shibasaki and coworkers developed a chiral lanthanum(III)/lithium alkoxide complex (La–Li complex). When used with 20 mol% of this complex moderate yields and low levels of enantioselectivity were obtained in reactions of acetophenone with oxygen-substituted pivaldehyde **8** (Scheme 3.3.5).[21]

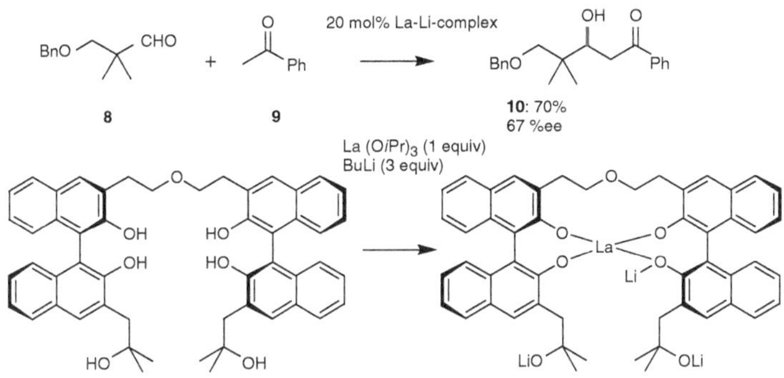

Scheme 3.3.5 Lanthanum/lithium-linked BINOL complex in direct enantioselective aldol addition

An improvement of these results was achieved by the synthesis and the application of a bridged BINOL–zinc complex. Hydroxyacetophenone was used as substrate in these direct aldol additions. High *syn*-diastereoselectivities and enantioselectivities were detected (Scheme 3.3.6).[22,23,24,25]

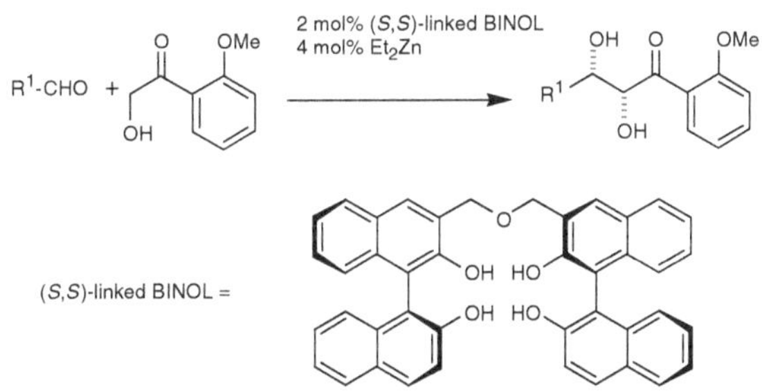

entry	R^1	yield (%)	*syn* / *anti*	ee (%) *syn* / *anti*
1	$Ph(CH_2)_2$	94	89 / 11	92 / 89
2	$CH_3(CH_2)_4$	88	88 / 12	95 / 91
3	*i*-Bu	84	84 / 16	93 / 87
4	$Me(C{=}O)(CH_2)_2$	91	93 / 7	95 / -
5	$BnO(CH_2)_2$	81	86 / 14	95 / 90
6	$BnOCH_2$	84	72 / 28	96 / 93
7	*i*-Pr	83	97 / 3	98 / -
8	Et_2C	92	96 / 4	99 / -
9	Cy	95	97 / 3	98 / -

Scheme 3.3.6 Diethylzinc-linked BINOL complex for the enantioselective direct aldol addition

This methodology was also used for the construction of defined configured tertiary alcohols. When used with 2-hydroxy-propiophenone the corresponding 1,2-diol-ketones were isolated with high yields. Moderate diastereoselectivities were detected (Scheme 3.3.7).[26]

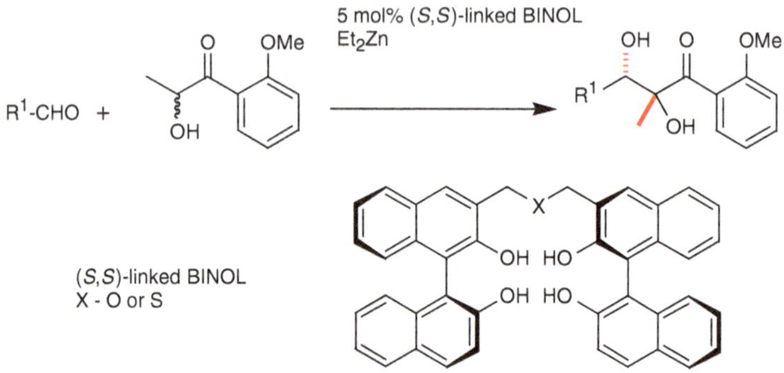

entry	R^1	catalyst	yield (%)	*syn* / *anti*	ee (%) *syn*	ee (%) *anti*
1	$PhCH_2CH_2$	O	97	38 / 62	96	87
2	$PhCH_2CH_2$	S	82	65 / 35	92	60
3	$Ph(CH_2)_3$	O	72	36 / 64	90	78
4	$Ph(CH_2)_3$	S	63	59 / 41	86	45
5	Et	O	88	29 / 71	86	68
6	Et	S	56	59 / 41	87	48
7	$PMBOCH_2CH_2$	O	89	41 / 59	95	86
8	$PMBOCH_2CH_2$	S	73	59 / 41	93	58
9	$BOMOCH_2CH_2$	O	92	31 / 69	97	87
10	$BOMOCH_2CH_2$	S	72	61 / 39	81	52
11	*i*-Pr	O	80	32 / 68	87	72
12	$BnOCH_2$	O	80	35 / 65	92	85

Scheme 3.3.7 Influence of the heteroatom on the diastereoselectivity of diethylzinc-linked BINOL-catalysed enantioselective direct aldol addition; X–O → 5 mol%; X–S → 10 mol%

The favoured formation of (*R*)-configured *syn*-product can best be explained with a transition state model shown in Fig. 3.3.1.

Trost and coworkers developed a dinuclear zinc complex. This complex was synthesized by the reaction of 1,2-aminoalcohol **11** and diethylzinc. By using 5 mol% of this catalyst high enantioselectivities were obtained (Scheme 3.3.8).[27]

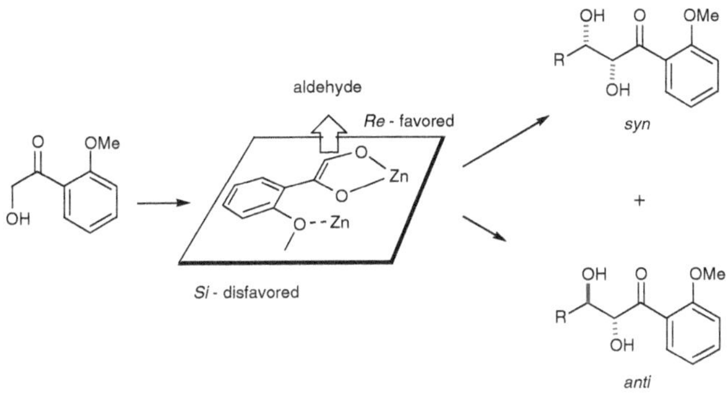

Fig. 3.3.1 Proposed transition state model

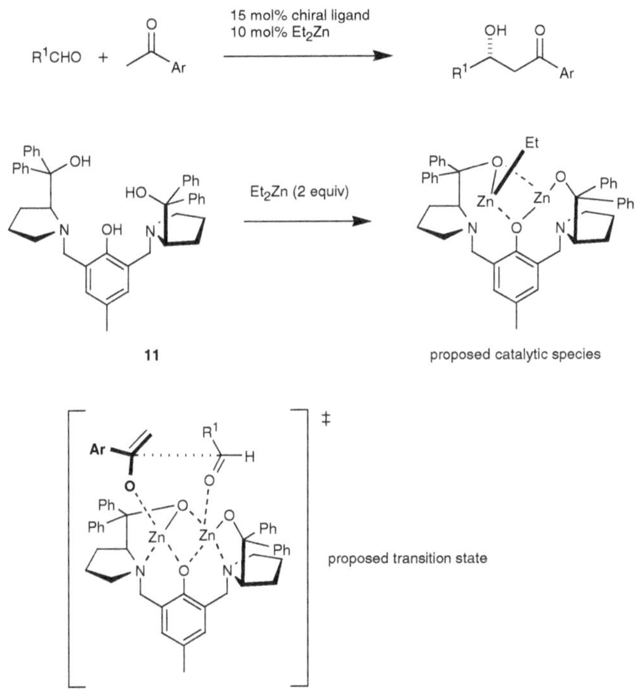

entry	R^1	Ar	yield (%)	ee (%)
1	*n*-Pr	Ph	33	56
2	*i*-Bu	Ph	49	68
3	*i*-Pr	Ph	62	98
4	*i*-Pr	2-furyl	66	97
5	*i*-Pr	2-MeO-C_6H_4	48	97
6	*i*-Pr	4-MeO-C_6H_4	36	98
7	Cy	Ph	60	98
8	TBSO$CH_2C(CH_3)_2$	Ph	61	93

Scheme 3.3.8 Direct asymmetric aldol reaction catalysed by dinuclear chiral zinc complex

The bifunctional catalyst is supposed to act with one zinc atom as a Lewis acid and the other zinc atom is acting as a Brønstedt base. The latter is generating the zinc enolate (proposed transition state, Scheme 3.3.8).

An improvement of yields and enantioselectivities is obtained by the application of the modified ligand **12** (Scheme 3.3.9).[28]

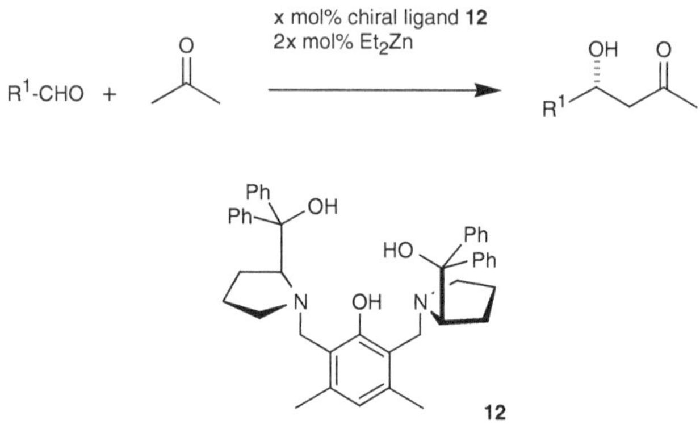

entry	R^1	mol% ligand **12**	yield (%)	ee (%)
1	Cy	10	89	92
2	*i*-Pr	10	89	91
3	*t*-Bu	10	72	94
4	*i*-Bu	10	59	84
5	*n*-Pr	10	69	89
6	Ph	5	78	83
7	4-NO_2-C_6H_4	5	62	78

Scheme 3.3.9 Improvement of yields and enantioselectivities by modification of ligand

Also, by modification of this binuclear zinc catalyst (ligand **11**) a more general application regarding the substrates used was possible. Unbranched hydroxyketones can be used as enolate components in these direct aldol additions. Moreover, the catalyst loading could be diminished, whereas yields and enantioselectivities were improved (Scheme 3.3.10).

For a similar catalyst system in enantioselective direct aldol additions of aryl ketones and aryl aldehydes see Li et al.[29] For applications of gallium(III) Lewis acids using ligand **11** and similar chiral aminoalcohol ligands in Mukaiyama aldol reactions see Li et al.[30]

Application of this methodology is given in the total synthesis of boronolide (Scheme 3.3.11)[31] and fostriecin (Scheme 3.3.12).[32]

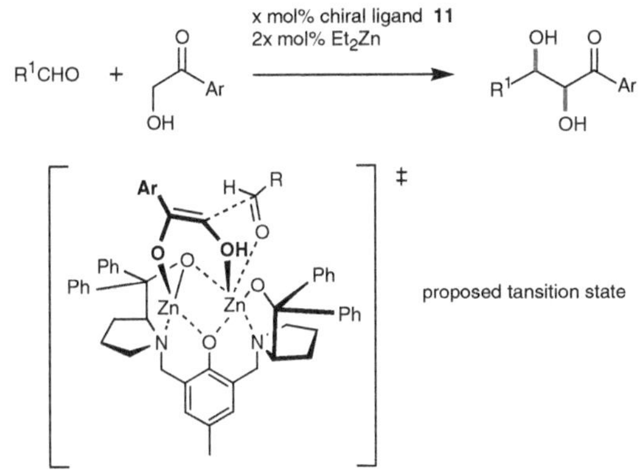

entry	R^1	Ar	mol% ligand	yield (%)	*syn* / *anti*	ee (%) *syn*
1	Cy	Ph	2.5	83	97 / 3	92
2	Cy	2-furyl	5	90	86 / 14	96
3	*i*-Pr	Ph	2.5	89	93 / 7	93
4	*i*-Bu	Ph	2.5	65	95 / 5	94
5	$PhCH_2CH_2$	Ph	2.5	78	90 / 10	91
6	$CH_3(CH_2)_6$	Ph	5	89	83 / 17	86
7	$H_2C=CH(CH_2)_8$	Ph	5	91	83 / 17	87

Scheme 3.3.10 Application of dinuclear zinc complex to aldol additions of α-hydroxyketones

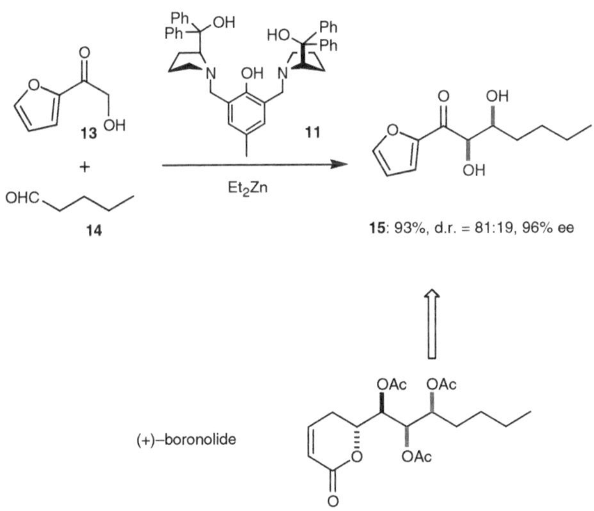

Scheme 3.3.11 Total synthesis of boronolide

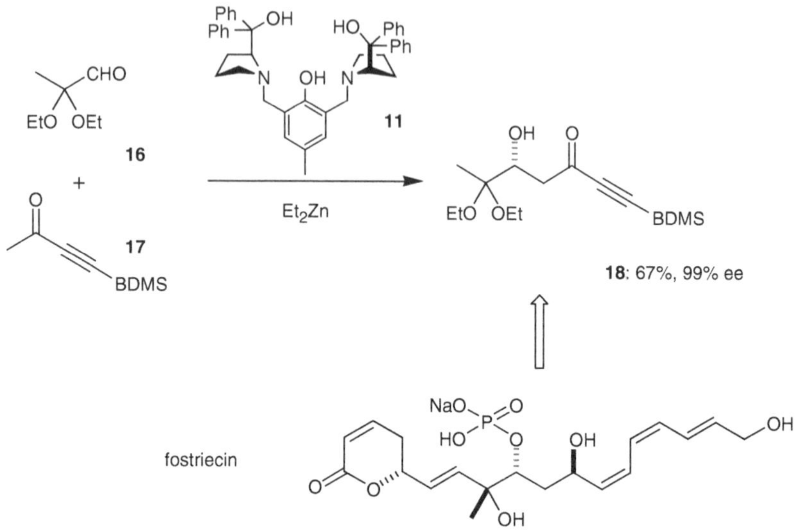

Scheme 3.3.12 Total synthesis of fostriecin

Noyori and coworkers described the synthesis and application of a chiral calcium complex. By deployment of this chiral calcium complex the aldol adducts were obtained in good yields and enantioselectivities. The reactivity of this complex is much higher compared with other complexes discussed in this chapter (Catalyst loading). However, high excesses of ketones are necessary to achieve the desired reaction (Scheme 3.3.13).[33]

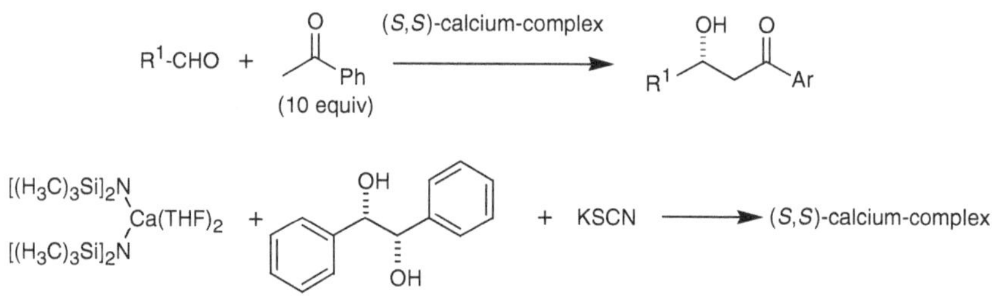

entry	R^1	mol% catalyst	yield (%)	ee (%)
1	*t*-Bu	1	79	82
2	$PhCH_2C(CH_3)_2$	3	75	87
3	$BnOCH_2C(CH_3)_2$	3	76	91
4	Cy	3	88	66

Scheme 3.3.13 Direct enantioselective aldol addition of aldehydes to acetophenone

Titanium(IV) complexes derived from BINOL and mandelic acid catalyse the direct and enantioselective aldol addition. 10 mol% of these complexes are sufficient enough for a full conversion. Interestingly, the enantioselectivity obtained in these transformations depends only on the chirality of the mandelic acid used. By application of (*R*)- or (*S*)-configured BINOL same results were obtained as in reactions performed out with racemic BINOL (Scheme 3.3.14).[34,35]

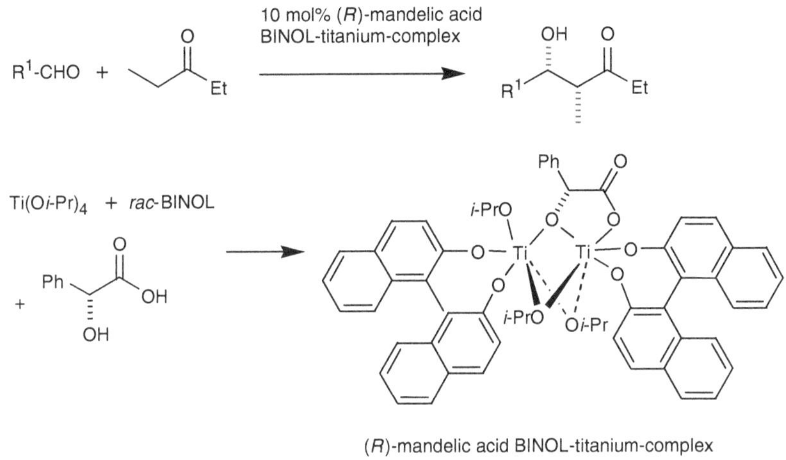

entry	R^1	yield (%)	*syn* / *anti*	ee (%) *syn*
1	Ph	85	91 / 9	91
2	*t*-Bu	71	88 / 12	93
3	PhC ≡ C	68	73 / 27	78
4	*i*-Pr	43	79 / 21	71
5	Et	78	72 / 28	74

Scheme 3.3.14 Mandelic acid as chiral source in direct *syn*-selective asymmetric aldol additions to diethyl ketone

The direct enantioselective aldol addition was extended to glycine derivates as ene components. Thus, an approach to chiral α-amino-β-hydroxy carboxylic esters is given. Miller and coworkers reported this transformation for the first time using chiral benzyl cinchonium chloride as catalyst.[36] Maruoka and coworkers developed a chiral phase transfer catalyst based on quaternary ammonium salts. This catalyst was employed in glycine Schiff base aldol addition. Results are given in Scheme 3.3.15.[37]

Also, Shibasaki and coworkers employed their own LLB methodology in these transformations. The aldol adducts were obtained in moderate diastereoselectivities and enantioselectivities.[38]

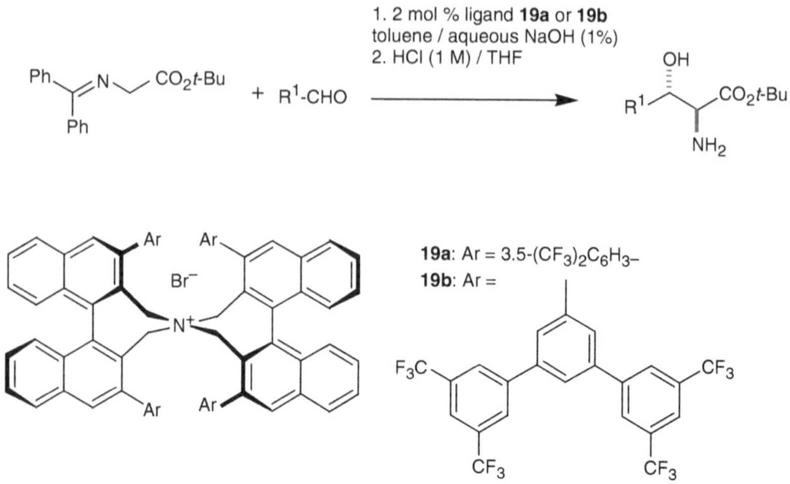

entry	R^1	catalyst	yield (%)	*anti* / *syn*	ee (%) *anti*
1	$PhCH_2CH_2$	**19a**	76	77 / 23	91
2	$CH_3(CH_2)_4$	**19a**	65	91 / 9	91
3	$TIPSOCH_2$	**19a**	72	95 / 5	98
4	$H_2C = CH(CH_2)_2$	**19a**	62	86 / 14	80
5	Me	**19b**	58	70 / 30	92
6	Cy	**19b**	40	74 / 26	95

Scheme 3.3.15 Glycine Schiff base aldol addition mediated by chiral quaternary ammonium salts

A copper-catalysed direct aldol addition of acetonitrile was published by Shibasaki and coworkers. They reacted a series of aldehydes with acetonitrile in the presence of 10 mol% of copper(I) *tert*-butoxide and chiral biphenylphosphanes. The expected β-hydroxynitriles were isolated in moderate to good enantioselectivities.[39]

A catalytic asymmetric synthesis of 11-deoxy-$PGF_{1\alpha}$ has been achieved by Shibasaki and coworkers using a direct tandem Michael-aldol reaction key step. This cascade reaction was efficiently promoted by the catalytic use of 5 mol% AlLibis[(*S*)-binaphthoxide] complex (ALB) to give the three-component coupling product **16** at room temperature in 97% ee and in 75% yield (Scheme 3.3.16).[40]

Several highly regio- as well as diastereoselective direct and catalytic aldol additions using titanium complexes were reported. When used with titanium(IV) complexes derived from BINOL or mandelic acid high regioselectivities were observed in direct aldol additions of hydroxyketones and aldehydes. Attack of the aldehydes was observed at the sterically more encumbered α-position of the ketones only.[41,42]

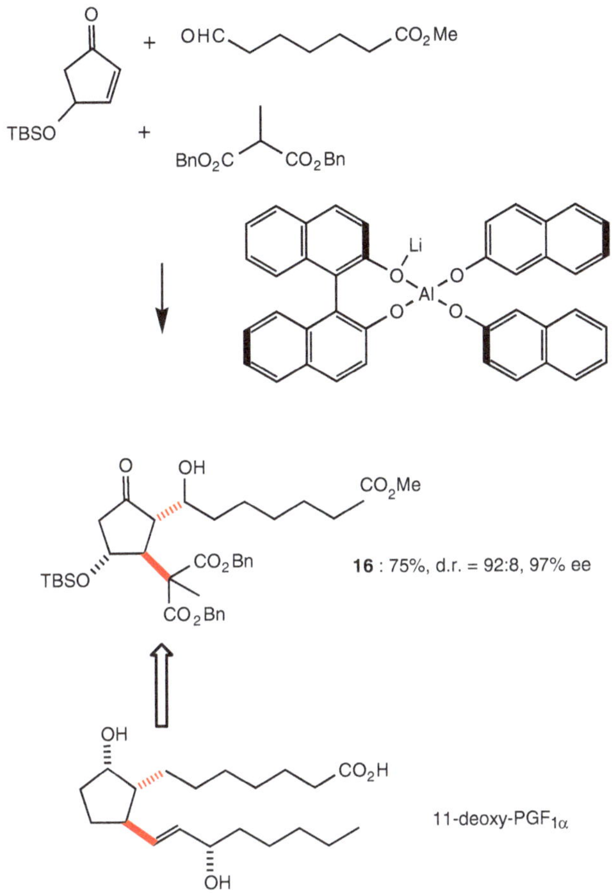

Scheme 3.3.16 Total synthesis of $PGF_{1\alpha}$ by using an ALB catalyst

For a direct and enantioselective Henry reaction in the presence of catalytic amounts of chiral lanthanide complexes see Saa et al.[43]

When used with chiral bis(oxazolinyl)phenyl–rhodium complexes in direct aldol reaction of ketones and aromatic aldehydes the corresponding β-hydroxyketones were obtained in high *anti*-diastereoselectivity and a good to high enantioselectivity up to 91% ee.[44]

References

1. Trost BM (2002) Acc Chem Res 35:695
2. Miyoshi N, Fukuma T, Wada M (1995) Chem Lett 999
3. Mahrwald R (1995) Chem Ber 128:919
4. Mahrwald R, Ziemer B, Troyanov S (2001) Tetrahedron Lett 42:6843

5. Mahrwald R, Gündogan B (1998) J Am Chem Soc 120:413
6. Yoshida Y, Hayashi R, Sumihara H, Tanabe Y (1997) Tetrahedron Lett 38:8727
7. Yoshida Y, Matsumoto N, Hamasaki R, Tanabe Y (1999) Tetrahedron Lett 40:4227
8. Tanabe Y, Matsumoto N, Higashi T, Misaki T, Ito T, Yamamoto M, Mitarai K, Nishii Y (2002) Tetrahedron 58:8269
9. Tanabe Y, Matsumoto N, Funakoshi S, Manta N (2001) Synlett 1959
10. Machajewski TD, Wong CH (2000) Angew Chem Int Ed Engl 39:1352
11. Dalko PI, Moisan L (2004) Angew Chem Int Ed Engl 43:5138
12. Berkessel A, Gröger H (2005) In: Asymmetric Organocatalysis, Wiley-VCH, Weinheim
13. Gaunt MJ, Johansson CCC, McNally M, Vo NT (2007) Drug Discovery Today 12:8
14. Alcaide B, Almendros P (2002) Eur J Org Chem 1595
15. Yamada YMA, Yoshikawa N, Sasai H, Shibasaki M (1997) Angew Chem Int Ed Engl 36:1871
16. Fujii K, Maki K, Kanai M, Shibasaki M (2003) Org Lett 5:733
17. Fujii K, Maki K, Kanai M, Shibasaki M (2003) Org Lett 5:733.
18. Sawada D, Kanai M, Shibasaki M (2000) J Am Chem Soc 122:10521
19. Yoshikawa N, Yamada YMA, Das J, Sasai H, Shibasaki M (1999) J Am Chem Soc 121:4168
20. Yoshikawa N, Kumagai N, Matsunaga S, Moll G, Ohshima T, Suzuki T, Shibasaki M (2001) J Am Chem Soc 123:2466
21. Yoshikawa N, Shibasaki M (2001) Tetrahedron 57:2569
22. Yoshikawa N, Kumagai N, Matsunaga S, Moll G, Ohshima T, Suzuki T, Shibasaki M (2001) J Am Chem Soc 123:2466
23. Yoshikawa N, Suzuki T, Shibasaki M (2002) J Org Chem 67:2556
24. Kumagai N, Matsunaga S, Kinoshita T, Harada S, Okada S, Sakamoto S, Yamaguchi K, Shibasaki M (2003) J Am Chem Soc 125:2169
25. Kumagai N, Matsunaga H, Yoshikawa N, Ohshima T, Shibasaki M (2001) Org Lett 3:1539
26. Kumagai N, Matsunaga S, Kinoshita T, Harada S, Okada S, Sakamoto S, Yamaguchi K, Shibasaki M (2003) J Am Chem Soc 125:2169.
27. Trost BM, Ito H (2000) J Am Chem Soc 122:12003
28. Trost BM, Silcoff ER, Ito H (2001) Org Lett 3:2497
29. Li H, Da CS, Xiao YH, Li X, Su YN (2008) J Org Chem 73:7398
30. (a) Li HJ, Tian HY, Wu YC, Chen YJ, Liu L, Wang D, Li CJ (2005) Adv Synth Catal 347:1247; (b) Li HJ, Tian HY, Chen YJ, Wang D, Li CJ (2002) Chem Commun 2994
31. Trost BM, Yeh YSC (2002) Org Lett 4:3513
32. Trost BM, Frederiksen MU, Papillon JPN, Harrington PE, Shin S, Shireman BT (2005) J Am Chem Soc 127:3666
33. Suzuki T, Yamagiwa N, Matsuo Y, Sakamoto S, Yamaguchi K, Shibasaki M, Noyori R (2001) Tetrahedron Lett 42:4669
34. Mahrwald R, Ziemer B (2002) Tetrahedron Lett 43:4459
35. Mahrwald R (2000) Org Lett 2:4011
36. Gasparski CM, Miller MJ (1991) Tetrahedron 47:5367
37. Ooi T, Taniguchi M, Kameda M, Maruoka K (2002) Angew Chem Int Ed 41:4542
38. Yoshikawa N, Shibasaki M (2002) Tetrahedron 58:8289
39. Suto Y, Tsujii R, Kanai M, Shibasaki M (2005) Org Lett 7:3757
40. Yamada KI, Arai T, Sasai H, Shibasaki M (1998) J Org Chem 63:3666
41. Mahrwald R, Schetter B (2006) Org Lett 8:281
42. Schetter B, Stosiek C, Ziemer B, Mahrwald R (2007) Appl Organomet Chem 21:139
43. Saa JM, Tur F, Gonzalez J, Vega M (2006) Tetrahedron: Asymm 17:99
44. Inoue H, Kikuchi M, Ito J-i, Nishiyama H (2008) Tetrahedron 64:493

3.4 Gold- and Rhodium-Catalysed Aldol Additions

In 1986 Ito and coworkers reported enantioselective aldol additions of methyl isocyanoacetate to aldehydes mediated by chiral ferrocenyl–gold complexes.[1,2,3] The corresponding *trans*-configured oxazolines were obtained with an excellent degree of enantioselectivity (Scheme 3.4.1). By hydrolysis the corresponding β-hydroxy-α-amino acids were obtained. This reaction has been extensively studied by Ito[4,5] as well as Togni and coworkers.[6,7,8,9,10,11,12]

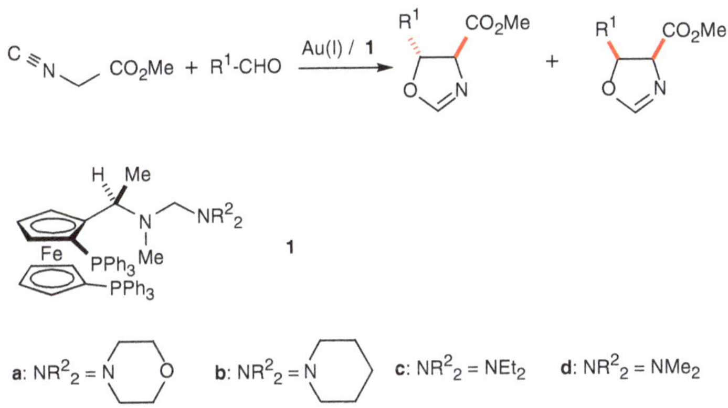

entry	R^1	yield (%)	*anti* / *syn*	ee (%, *anti*)
1	Ph	93	95 / 5	95
2	2-Me-C_6H_4	98	96 / 4	95
3	4-Cl-C_6H_4	97	94 / 6	92
4	4-NO_2-C_6H_4	80	83 / 17	86
5	Me	99	89 / 11	89
6	*i*-Bu	99	96 / 4	87

Scheme 3.4.1 Ferrocenyl–gold complexes in enantioselective aldol addition of aldehydes to isocyanoacetates; NR^2_2:**a**

R. Mahrwald, *Aldol Reactions*,

This methodology was extended to the use of α-branched isocyanocarboxylates. Thus, the authors were able to isolate the *anti*-configured adducts with a high degree of enantioselectivity (Scheme 3.4.2).[13]

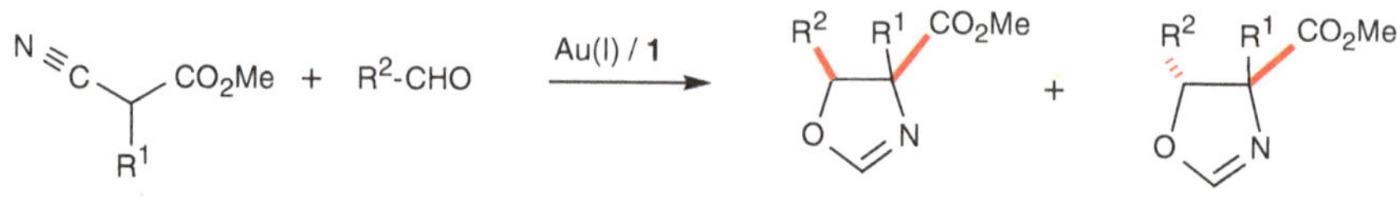

entry	R^1	R^2	ligand	yield (%)	*syn* / *anti*	ee (%)*anti*	ee (%) *syn*
1	Me	Ph	a	97	93 / 7	94	53
2	Me	Ph	b	92	88 /12	90	5
3	Me	Ph	c	90	77 / 23	82	26
4	Me	Ph	d	95	82/18	92	44
5	*i*-Pr	Ph	a	86	62 / 38	88	17
6	*i*-Pr	Ph	b	86	54 / 46	92	28
7	*i*-Pr	Ph	c	87	52 / 48	85	42
8	*i*-Pr	Ph	d	95	50 / 50	88	48
9	Me	Me	a	86	56 / 44	86	54
10	Me	Me	b	94	44 / 56	44	6
11	Me	Me	d	100	38 / 62	46	49
12	Et	Me	a	92	54 / 46	87	66
13	*i*-Pr	Me	a	100	24 / 76	26	51
14	*i*-Pr	Me	b	100	22 / 78	35	23

Scheme 3.4.2 Enantioselective aldol addition of aldehydes to α-branched isocyanoacetates

An aldol addition catalysed by a chiral diphosphanyl–rhodium(I) complex yields nitriles containing a quaternary chiral carbon centres at the α-position (Scheme 3.4.3).[14] Complete conversion was obtained with 1 mol% of a rhodium catalyst containing the chiral (*S*,*S*)-(*R*,*R*)-TRAP ligand (Scheme 3.4.3).

The observed stereochemistry at the α-position of the aldol product suggests that (*S*,*S*)-(*R*,*R*)-TRAP ligand on the rhodium complex differentiates between the steric bulkiness of the α-methyl and the ester group of 2-cyanopropanoates.

The preferential formation of *anti*-configured aldol adducts in entries 11–15 (Scheme 3.4.3) suggests that this reaction proceeds through the antiperiplanar transition state **A** where steric interactions between the aldehyde substituent (R^2)

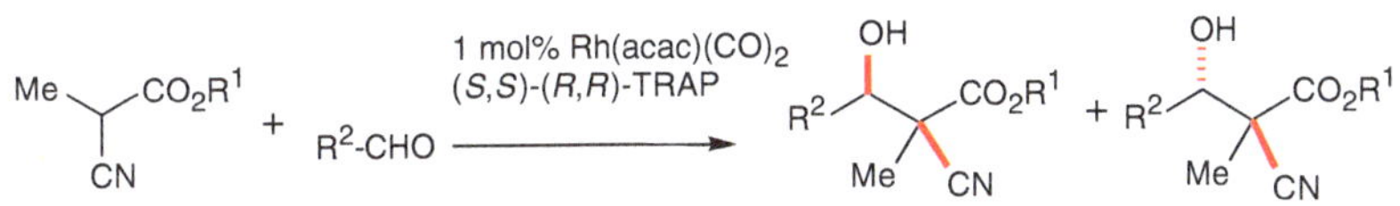

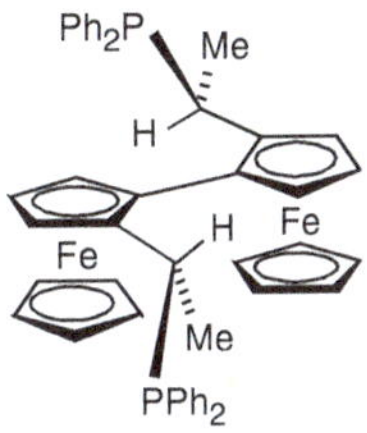

(*S*,*S*)-(*R*,*R*)-TRAP

entry	R^1	R^2	yield (%)	*anti* / *syn*	ee (%)	
1	Et	H	84	-	60	
2	Me	H	67	-	35	
3	Et	H	85	-	74	
4	*i*-Pr	H	86	-	78	
5	*n*-Bu	H	80	-	82	
6	CH*i*-Pr$_2$	H	82	-	91	
7	CHn-Bu$_2$	H	86	-	93	
8	CHPh$_2$	H	96	-	87	
9	Et	Me	63	45 / 55	31	23
10	*i*-Pr	Me	61	47 / 53	55	50
11	CH*i*-Pr$_2$	Me	67	81 / 19	86	33
12	CH*i*-Pr$_2$	Et	76	75 / 25	57	10
13	CH*i*-Pr$_2$	EtO$_2$C	88	68 / 32	91	63

Scheme 3.4.3 TRAP as chiral ligand for rhodium-catalysed addition of aldehydes to 2-cyanopropionates

and the bulky diisopropylmethyl ester group (R^1) are avoided (Scheme 3.4.4). The synclinal transition state **C** appears to be less favourable than **D** owing to the steric interaction between R^2 and one of the phenyl groups of (*S*,*S*)-(*R*,*R*)-TRAP. The low diastereoselectivity of the reactions of 2-cyanopropionates with sterically less-demanding ester groups (entries 9 and 10, Scheme 3.4.3) may be due to the lower steric repulsion between R^2 and the ester group.

Later on Ito and coworkers have demonstrated the same reactions using corresponding chiral silver(I) complexes.[15]

An extension of these works has recently been reported by Willis. Enantioselective aldol additions of aromatic aldehydes to isothiocyanates give access to chiral β-hydroxy-α-amino acids.[16]

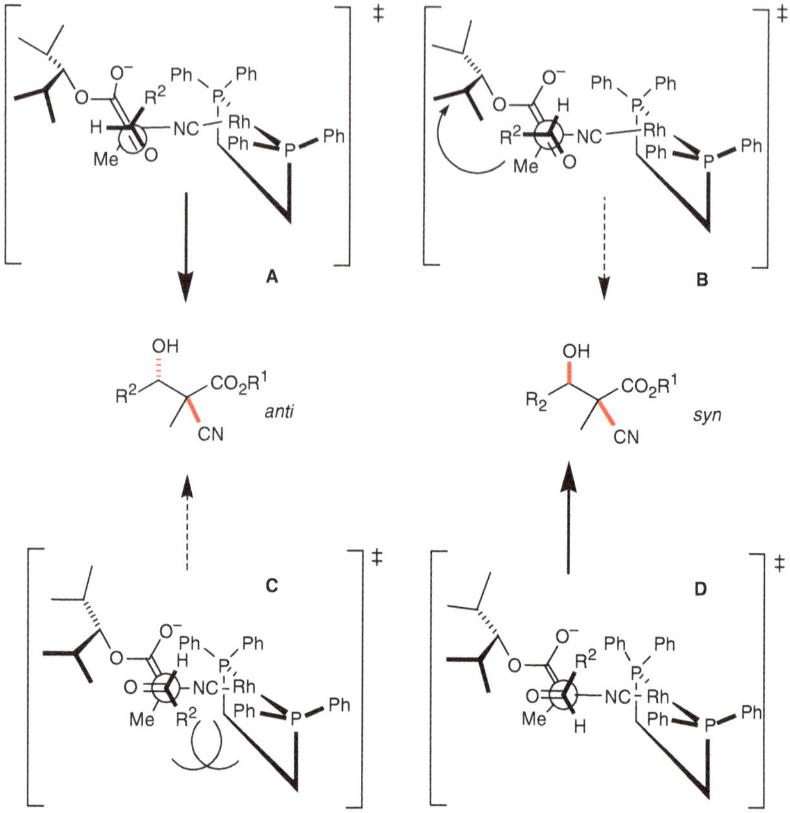

Scheme 3.4.4 Transition state models for the stereochemical outcome of the rhodium-catalysed addition of 2-cyanopropionates to aldehydes

Richards and Stark reported a direct palladium-catalysed aldol addition of methyl isocyanoacetate.[17] A direct enantioselective rhodium-catalysed aldol addition was described by Nishiyama and coworkers.[18] In reactions of cyclopentanone, cyclohexanone or acetone with activated benzaldehydes moderate to good enantioselectivities were detected. For direct and regioselective rhodium-catalysed aldol additions see also Murakami et al.[19]

For reductive aldol additions in the presence of chiral rhodium complexes, see publications in reference.[20] In these transformations unsaturated ketones were reacted with aldehydes in the presence of chiral rhodium catalysts. The unsaturated ketones were reduced to give the expected aldol adducts in reactions with aldehydes. By a fine-tuning of ligands and additives the undesired 1.4-reduction can be completely suppressed.

In reactions with unsaturated nitriles, esters or ketones in the presence of rhodium complexes a reductive aldol-Tishchenko reaction is observed (Scheme 3.4.5).[21] The corresponding esters were obtained in high yields.

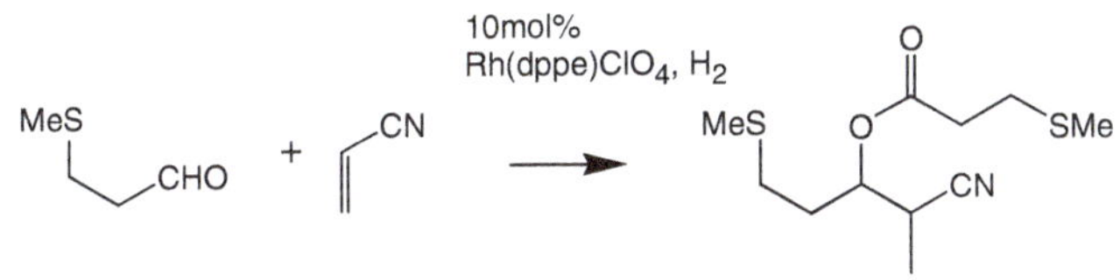

Scheme 3.4.5 Reductive aldol-Tishchenko reaction in the presence of rhodium catalyst

Similar reactions in the presence of hydrogen give access to all *syn*-configured stereotriads. Krische and coworkers demonstrated this by reactions of a series of optically active aldehydes. When used with α-aminoaldehydes an access to defined-configured nitrogen-substituted stereotriads is given (Scheme 3.4.6).[22]

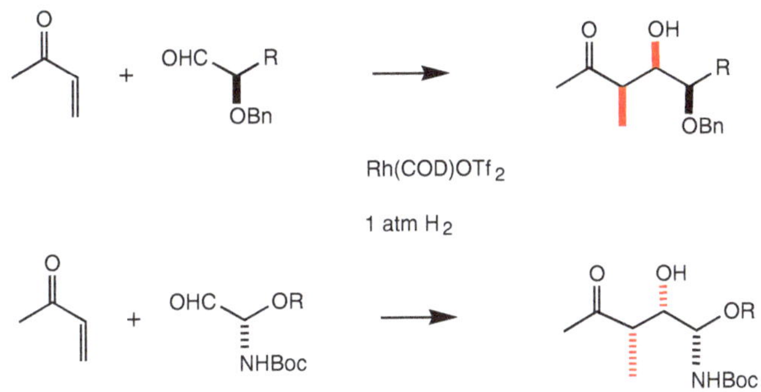

Scheme 3.4.6 Aldol additions catalysed by rhodium complexes in hydrogen atmosphere

A reductive enantioselective aldol addition in the presence of BINAP/rhodium complexes was reported by Morken and coworkers. Aldol adducts of aldehydes to acrylates were obtained in moderate diastereoselectivities, but with high degrees of enantioselectivities.[23,24,25]

References

1. Ito Y, Sawamura M, Hayashi T (1986) J Am Chem Soc 108:6405
2. Ito Y, Sawamura M, Hayashi T (1987) Tetrahedron Lett 28:6215
3. Hayashi T, Sawamura M, Ito Y (1992) Tetrahedron 48:1999
4. Sawamura M, Ito Y (1992) Chem Rev 92:857
5. Sawamura M, Ito Y (1993) In: Catalytic Asymmetric Synthesis, Ojima I (ed). Wiley-VCH, New York, p 367
6. Pastor SD, Togni A (1989) J Am Chem Soc 111:2333
7. Togni A, Pastor SD (1989) Helv Chim Acta 72:1038
8. Togni A, Pastor SD, Rihs G (1990) J Organomet Chem 381:C21
9. Togni A, Pastor SD (1990) J Org Chem 55:1649
10. Pastor SD, Togni A (1991) Helv Chim Acta 74:905

11. Pastor SD, Togni A (1990) Tetrahedron Lett 31:839
12. Togni A, Pastor SD, Rihs G (1989) Helv Chim Acta 72:1471
13. Ito Y, Sawamura M, Shirawaka E, Hayashizaki K, Hayashi T (1988) Tetrahedron 44:5253
14. Kuwano R, Miyazaki H, Ito Y (2000) J Organomet Chem 603:18
15. Sawamura M, Hamashima H, Ito Y (1990) J Org Chem 55:5935
16. Willis MC, Cutting GA, Piccio VJD, Durbin MJ, John MP (2005) Angew Chem Int Ed 44:1543
17. Stark MA, Richards CJ (1997) Tetrahedron Lett 38:5881
18. Inoue H, Kikuchi M, Ito JI, Nishiyama H (2008) Tetrahedron 64:493
19. Murakami K, Ohmiya H, Yorimitsu H, Oshima K (2008) Tetrahedron Lett 49:2388
20. (a) Huddleston RR, Krische MJ (2003) Synlett 12; (b) Jang HY, Krische M (2004) Eur J Org Chem 3953; (c) Jang HY, Krische M (2004) J Acc Chem Res 37:653; (d) Chiu P (2004) Synthesis 2210
21. Wills, MC, Woodward RL (2005) J Am Chem Soc 127:18012
22. Jung CK, Krische MJ (2006) J Am Chem Soc 128:17051
23. Taylor SJ, Duffey MO, Morken JP (2000) J Am Chem Soc 122:4528
24. Zhao CX, Duffey MO, Taylor SJ, Morken JP (2001) Org Lett 3:1829
25. Taylor SJ, Morken JP (1999) J Am Chem Soc 121:12202

3.5 Antibody-Catalysed Aldol Addition

Antibodies are able to catalyse a great number of organic transformations.[1] Antibodies derived from aldolases are able to catalyse enantioselective aldol additions.[2] These reactions proceed via the known lysin-enamine mechanism of class I aldolases.[3] Synthetic working chemists designed antibodies with substrate specificity that can differ from those of natural occurring models. Lerner and Barbas developed several useful antibodies and tested them in enantioselective aldol additions.[4,5,6,7,8,9,10] Results of these investigations are shown in Scheme 3.5.1.

entry	R^1	antibody	configuration	ee (%)
1	4-*i*-PrCONH-C_6H_4	38C2	(*S*)	>99
2	4-NO_2-C_6H_4	38C2	(*S*)	98
3	4-NO_2-C_6H_4CH = CH	38C2	(*S*)	99
4	4-*i*-PrCONH-C_6H_4	93F3	(*R*)	>99
5	4-NO_2-C_6H_4	93F3	(*R*)	95

Scheme 3.5.1 Antibody-catalysed aldol reactions of aromatic aldehydes to acetone

The extent of enantiocontrol is usually very high for most reactions. In several cases both enantiomers could be accessed through the use of different antibodies (compare the use of antibodies 38C2 and 93F3 in Scheme 3.5.1). When used with unsymmetrical methyl ketones high regioselectivities were observed.[11]

More recently, Gouverneur and coworkers have published studies on regioselectivities observed in antibodies 38C2- and 84G3-mediated aldol additions.[12,13] When used with α- or β-heteroatom-substituted methyl ketones high regioselectivities were obtained.[14] Results are shown in Scheme 3.5.2.

R. Mahrwald, *Aldol Reactions*,

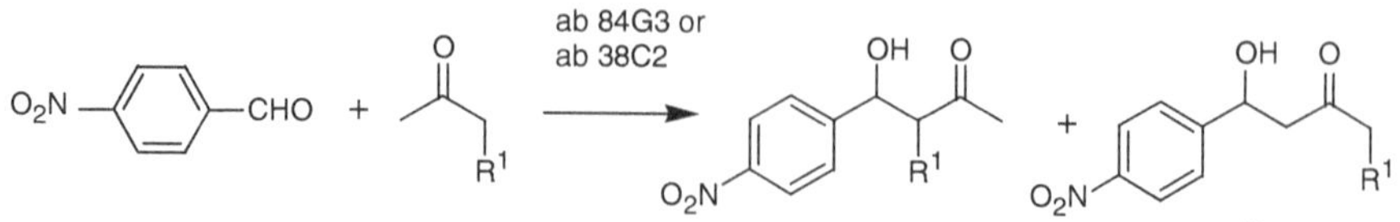

B

entry	R^1	antibody	conversion (%)	ratio A / B	ee (%)
1	OMe	ab84G3	76	0 / 100	98
2	OMe	ab38C2	35	100 / 0	n.r.
3	SMe	ab84G3	56	2 / 98	97
4	SMe	ab38C2	11	99 / 1	n.r.
5	Cl	ab84G3	90	5 / 95	n.r.
6	F	ab84G3	10	30 / 70	99
7	OH	ab84G3	n.r.	50 / 50	95

Scheme 3.5.2 Regioselective antibody-catalysed aldol additions to unsymmetrical ketones

An application of this aldol methodology was found in the total synthesis of epothilones. The starting chiral intermediate **2** was synthesized by an antibody-catalysed aldol addition of acetone and aldehyde **1** (Scheme 3.5.3). Furthermore, this antibody methodology was used in enantioselective retro-aldol addition of racemic aldol adducts to obtain the optically pure aldol adducts.[15]

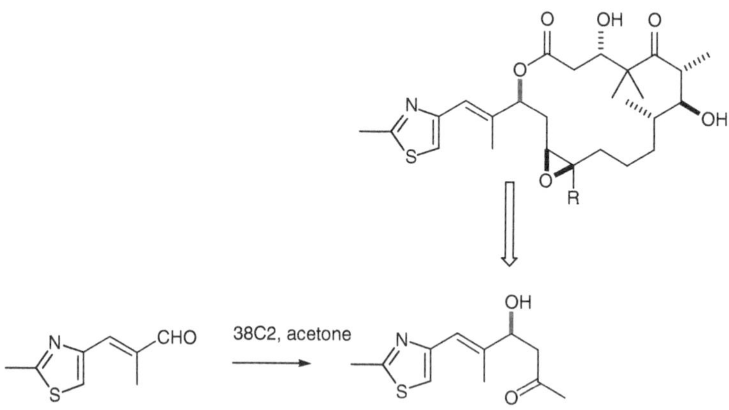

1

2: 40%, 96% ee

Scheme 3.5.3 Total synthesis of epothilone

References

1. Keinan E (2005) In: Catalytic Antibodies, Wiley-VCH, Weinheim
2. Wagner J, Lerner R, Barbas CF III (1995) Science 270:1797

3. Heine A, DeSantis G, Luz JG, Mitchell M, Wong CH, Wilson IA (2001) Science 294:369
4. Wagner J, Lerner RA, Barbas CF III (1995) Science 270:1797
5. Hoffmann T, Zhong G, List B, Shabat D, Anderson J, Gramatikova S, Lerner RA, Barbas CF III (1998)) J Am Chem Soc 120:2768
6. Barbas CF III, Heine A, Zhong G, Hoffmann T, Gramatikova S, Bjornestedt R, List B, Anderson J, Stura EA, Wilson IA, Lerner RA (1997) Science 278:2085
7. Zhong G, Hoffmann T, Lerner, RA, Danishefsky S, Barbas CF III (1997) J Am Chem Soc 119:8131
8. List B, Lerner RA, Barbas CF III (1999) Org Lett 1:59
9. List B, Lerner RA, Barbas CF III (1999) Org Lett 1:353
10. Karlstrom A, Zhong G, Rader C, Larsen NA, Heine A, Fuller R, List B, Tanaka F, Wilson IA, Barbas CF III, Lerner RA (2000) Proc Natl Acad Sci USA 97:3878
11. Zhong G, Lerner RA, Barbas CF III (1999) Angew Chem Int Ed 38:3738
12. Maggiotti V, Resmini M, Gouverneur W (2002) Angew Chem Int Ed 41:1012
13. Maggiotti V, Wong JB, Razet R, Cowley AR, Gouverneur V (2002) Tetrahedron: Asymm 13:1789
14. Maggiotti V, Bahmanyar S, Reiter M, Resmini M, Houk KN, Gouverneur V (2004) Tetrahedron 60:619
15. Sinha SC, Barbas CFIII, Lerner RA (1998) Proc Natl Acad Sci USA 95:14604

3.6 Enzyme-Catalysed Aldol Addition

Enzymes – in particular aldolases – are not only a supplement to classical methodologies, but also useful tools in many fields of aldol additions where classical synthetic procedures fail. This is the area of the synthesis of amino acids[1,2] and carbohydrates.[3,4,5,6,7,8] in the traditional sense. But more and more aldolases are used in C–C bond formation processes. Examples in natural product synthesis are given in Schetter and Mahrwald.[9] Moreover, the combination of these two different synthetic strategies – the enzymatic route and the classical organic synthetic route – represent a valuable tool for the construction of defined stereogenic centres of natural products.[10]

Natural occurring aldolases are classified by their mode of action into two main groups: class I and class II aldolases. The accepted reaction mechanism of these two groups is shown below. Class I aldolases bind substrates via imine-enamine formation with a lysine residue of the active site of the enzyme. This step initiates the C–C bond formation process with an aldehyde and subsequent hydrolysis sets the aldol adduct free (Scheme 3.6.1).

Scheme 3.6.1 Reaction mechanism for class I aldolases

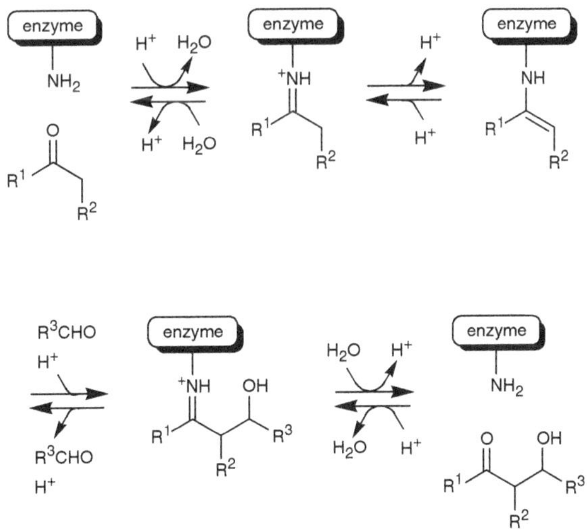

R. Mahrwald, *Aldol Reactions*,

Class II aldolases work with zinc ions, activating the enolate component, while at the same time a tyrosine residue from the adjoining subunit assists in the activation of the incoming aldehyde (Scheme 3.6.2).

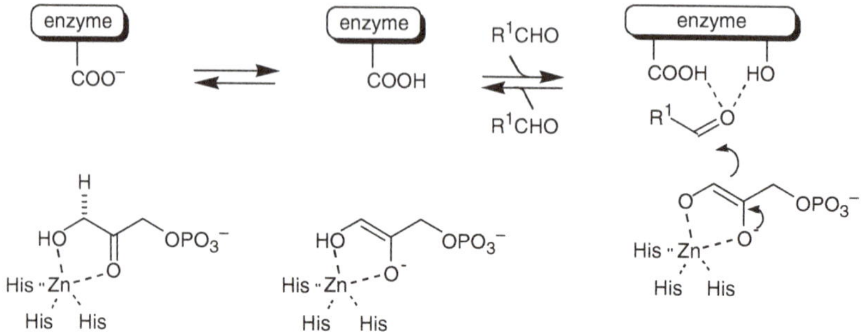

Scheme 3.6.2 Reaction mechanism for class II aldolases

Due to the growing interest of application of aldolases in organic synthesis a great number of different aldolases are available now. For that reason only a selected number of transformations with regard to their application in organic synthetic chemistry will be discussed here. For further studies read the very comprehensive overviews of this development.[11]

Pyruvate Aldolases

N-Acetylneuraminic acid aldolase catalyses the reversible addition of pyruvate **2** to *N*-acetyl-D-mannosamine **1** (Scheme 3.6.3).[12]

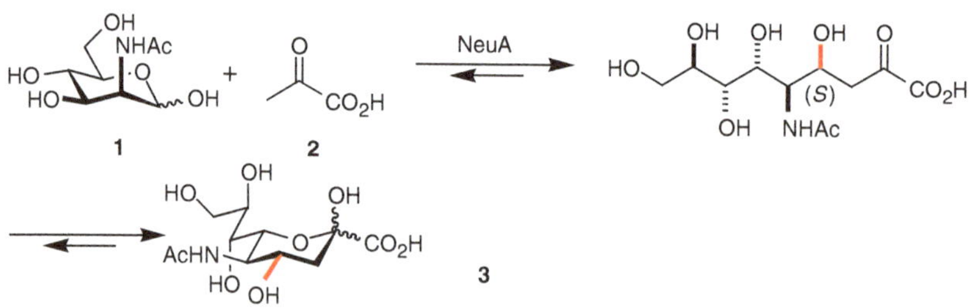

Scheme 3.6.3 *N*-Acetylneuraminic acid aldolase (NeuA)-catalysed aldol addition of pyruvate to *N*-acetyl-D-mannosamine

Neuraminic acid is an important precursor to *Zanamivir* produced for treatment against influenza infections. The industrial *N*-acetylneuraminic acid aldolase-mediated production of neuraminic acid **3** on a multiton-scale represents a benchmark of application of aldolases in industrial processes (Scheme 3.6.4).[13]

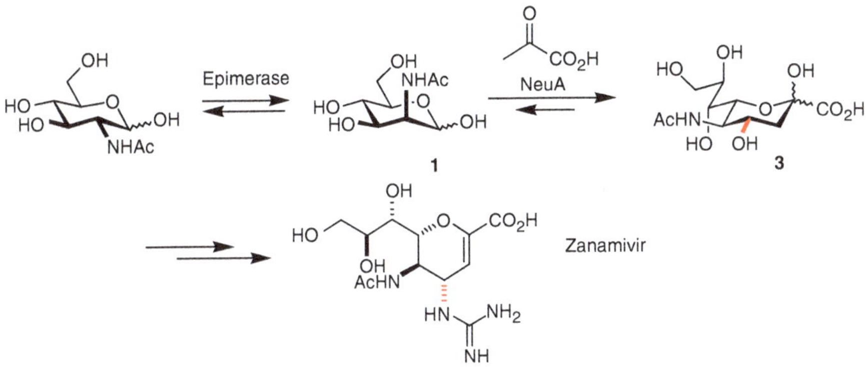

Scheme 3.6.4 Industrial application of NeuA in the synthesis of Zanamivir

A very recent example demonstrates the efficiency of 'direct evolution' in enzyme design.[14,15,16] *N*-Acetylneuraminic acid lyase (NAL) exhibits poor facial selectivity during C–C bond formation, and as such, its utility as a catalyst for the use in synthetic chemistry is limited. The group of Berry and Nelson was able to engineer the stereochemical course of NAL-catalysed C–C bond formation process by directed evolution and thus they were able to remove this limitation.[17]

Another example illustrates the synthesis of alkaloid intermediates. The starting neuraminic acid intermediate **4** was obtained by an enzymatic aldol reaction with pyruvate **2** (Scheme 3.6.5).[18]

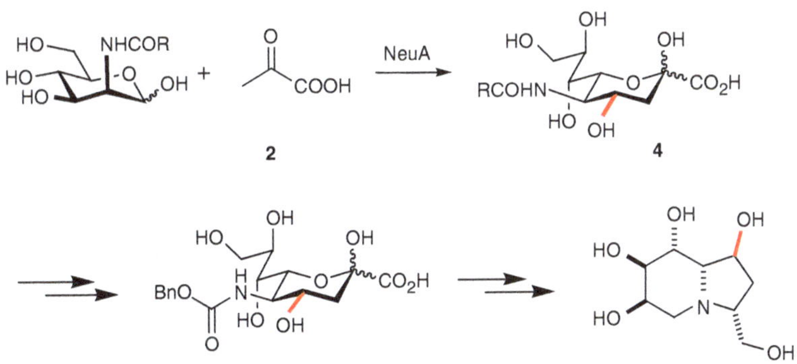

Scheme 3.6.5 NeuA-assisted synthesis of Castanospermine-analogs

2-Keto-3-deoxy-6-phospho-D-gluconate aldolase belongs to class I aldolases and has a rather broad substrate tolerance for polar and short chain aldehydes. (*S*)-configuration at C-4 and extremely high *anti*-diastereoselectivities were observed (Scheme 3.6.6).[19,20]

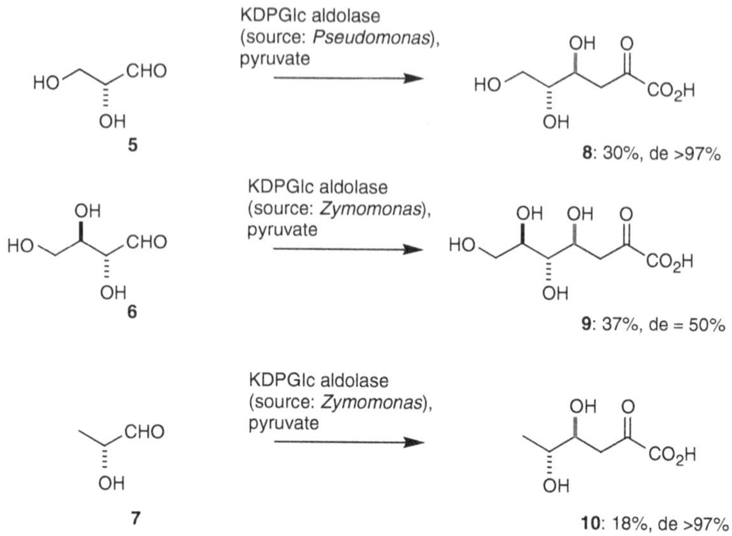

Scheme 3.6.6 Substrate tolerance of 2-keto-3-deoxy-6-phospho-D-gluconate aldolase[21]

Dihydroxyacetone Aldolases

The above discussed pyruvate aldol additions with *N*-acetylneuraminic acid aldolase and 2-keto-3-deoxy-6-phospho-D-gluconate aldolase create only one new stereoegenic centre. The products are derived from 'acetate' aldol addition. In contrast to that and with regard to chiral 'economy' the dihydroxyacetone phosphate aldolase-catalysed aldol addition represents the more efficient way. Dihydroxyacetone phosphate aldolases catalyse in vivo the addition of dihydroxy-acetone phosphate to D-glyceraldehyde-phosphate. As a result of this process two new stereogenic centres are formed with an extremely high selectivity. Four types of aldolases with distinct stereospecifity and with a broad substrate tolerance exist.[22,23,24] These are D-fructose 1,6-bisphosphate aldolase (FruA), D-tagatose 1,6-bisphosphate aldolase (TagA), L-rhamnulose 1,6-bisphosphate aldolase (RhuA) and L-fuculose 1,6-bisphosphate aldolase (FucA) (Scheme 3.6.7).

For application of this aldolase-catalysed aldol additions see references FruA,[25] TagA,[26] RhuA[27] and FucA.[28] Results of different stereoselective working modes of RhuA and FucA to several different aldehydes are given in Scheme 3.6.8.[29,30] For application of this methodology in total synthesis of nojirimycines see Ziegler et al.[31]

Based on this comfortable situation dihydroxyacetone phosphate aldolases have proved to be exceptionally powerful tools in asymmetric synthesis, particularly in stereocontrolled synthesis of polyoxygenated compounds. This fact is reflected by numerous transformations with several aldehydes as substrates on a preparative scale.[32]

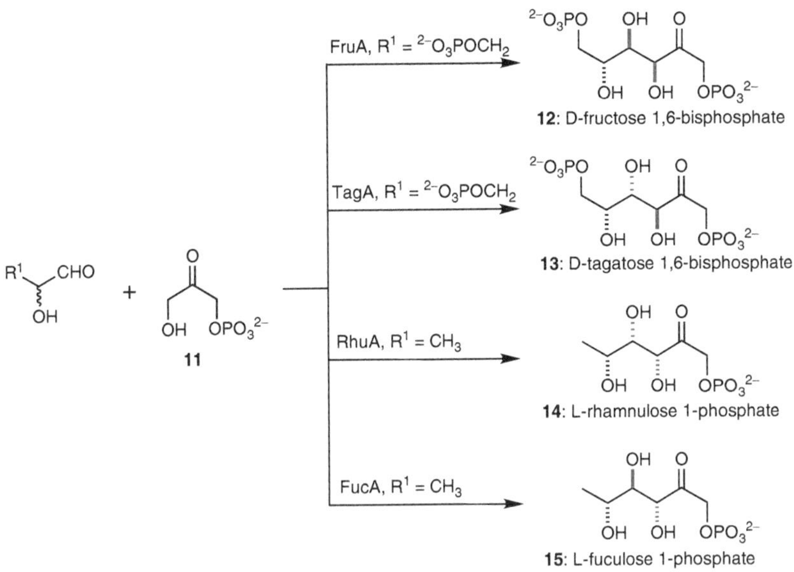

Scheme 3.6.7 Aldol reactions catalysed in vivo by the four stereo-complimentary dihydroxyacetone phosphate-dependent aldolases

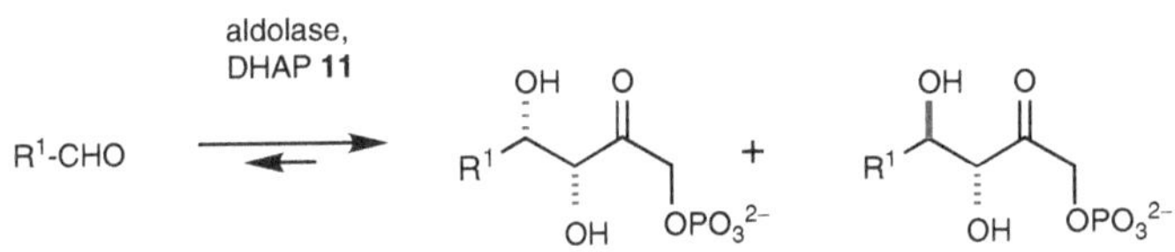

		RhuA		FucA	
entry	R^1	*syn* / *anti*	yield (%)	*syn* / *anti*	yield (%)
1	L-CH_2OH-Me	>97 / 3	95	<3 / 97	83
2	CH_2OH	>97 / 3	82	<3 / 97	85
3	D-CHOH-CH_2OH	>97 / 3	84	<3 / 97	82
4	CH_2-CH_2OH	>97 / 3	73	<3 / 97	78
5	CHOH-CH_2OMe	>97 / 3	77	<3 / 97	83
6	CHOH-CH_2N_3	>97 / 3	97	<3 / 97	80
7	CHOH-CH_2F	>97 / 3	95	<3 / 97	86
8	H	-	81	-	73
9	Me	69 / 31	84	5 / 95	54
10	$CHMe_2$	97 / 3	88	30 / 70	58

Scheme 3.6.8 Substrate tolerance of L-rhamnulose 1-phosphate (RhuA) and L-fuculose 1-phosphate aldolases (FucA)

Applications of chemoenzymatic reactions were reported for the synthesis of D-olivose (FruA),[33,34] L-fucose and derivatives (FucA),[35,36] thiosugars,[37,38] stereodivergent synthesis of 1-deoxy azasugars,[39,40,41,42,43,44] australine and epiaustraline (Frua),[45] brevicomin (FruA),[46] syringolide (FruA),[47] pentamycin (FruA),[48] aspillicin,[49] polyhydroxylated pyrrolidines (FruA),[50,51,52] fagomine (FruA),[53,54] polyhydroxylated piperidines (FruA),[55,56,57,58] polyhydroxylated piperidines (FucA),[59] azepanes,[60] atorvastatin[61] and vaiolamine.[62]

A very interesting application of this transformation is the tandem ozonolysis/aldolization process of suitable substituted olefins. Two subsequent aldol additions to dihydroxyacetone phosphate lead to higher carbon disaccharides (Scheme 3.6.9).[63]

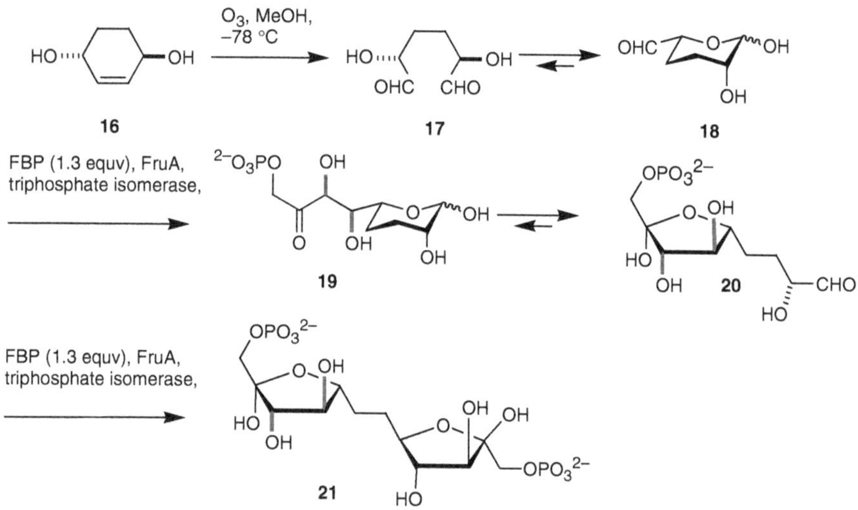

Scheme 3.6.9 Enzymatic aldol additions to disaccharide mimetics

Fessner and coworkers were able to extend this transformation to other substrates as illustrated in Scheme 3.6.10. Starting with racemic diols **22–26** only a single diastereoisomer can be obtained. The formation of pyranoide or furanoide structures depends on the substitution pattern of hydroxyl group (compare **27** and **28** in Scheme 3.6.10).

2-Deoxy-D-Ribose 5-Phosphate Aldolases

An approach to deoxypentoses is given by aldol additions catalysed by 2-deoxy-D-ribose 5-phosphate aldolases. The 2-deoxy-D-ribose 5-phosphate aldolase catalyses the aldol addition of acetaldehyde **34** to D-glyceraldehyde phosphate **32** and

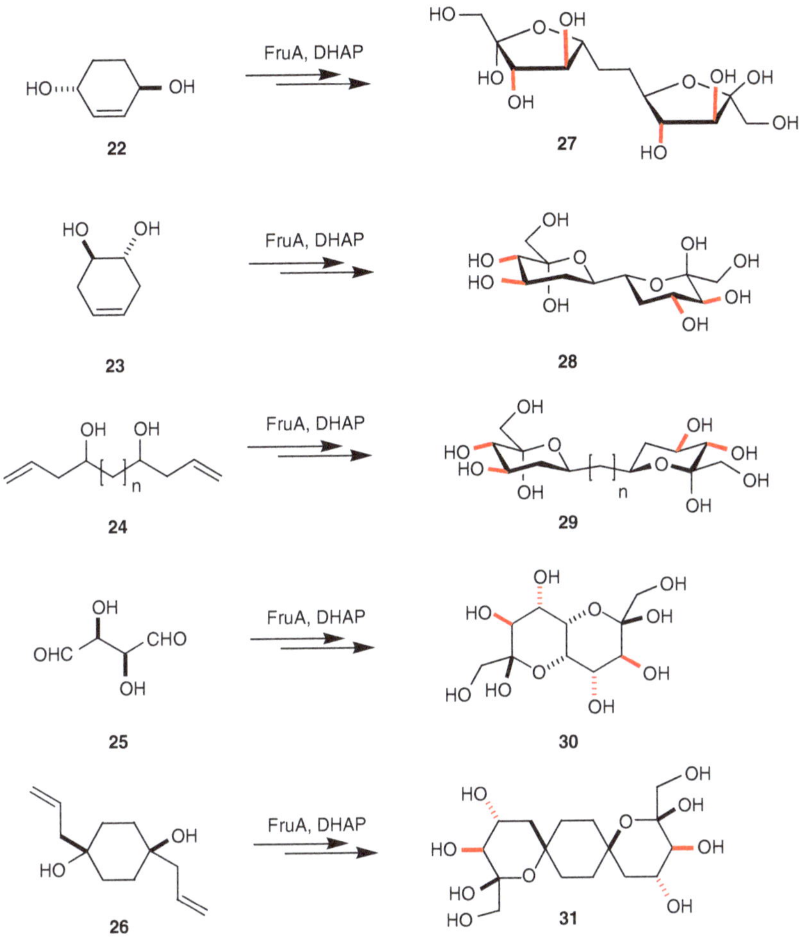

Scheme 3.6.10 Range of the synthetic possibilities of enzyme-catalysed aldol reactions in carbohydrate synthesis

belongs to class I aldolases. Thus, an approach to chiral β-hydroxyaldehydes is given. Schemes 3.6.11 and 3.6.12 give an overview of acceptance of non-natural aldol donors.[64,65] Selected examples are shown in Scheme 3.6.11. The β-carbon atom, which bears the hydroxy group is (*S*)-configured in every case. For an unprecedented asymmetric aldol addition of three C2-aldehydes catalysed by 2-deoxy-D-ribose 5-phosphate aldolase see Gijsen and Wong.[66]

The main field of application of enzyme-catalysed aldol additions has proved to be the de novo syntheses of carbohydrates. Nevertheless, there are some examples of stereoselective total synthesis of polyketides by enzyme-catalysed aldol additions. In particular, the high stereoselectivity of aldolases renders them very valuable in catalytic C–C bond formation processes. Some examples will demonstrate the usefulness of these biocatalysts.

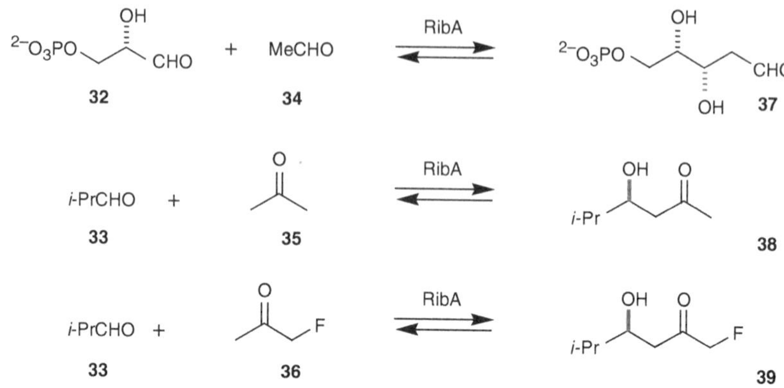

Scheme 3.6.11 Enzyme-catalysed cross-aldol additions

Scheme 3.6.12 Substrate tolerance of deoxy-D-ribose 5-phosphate aldolase

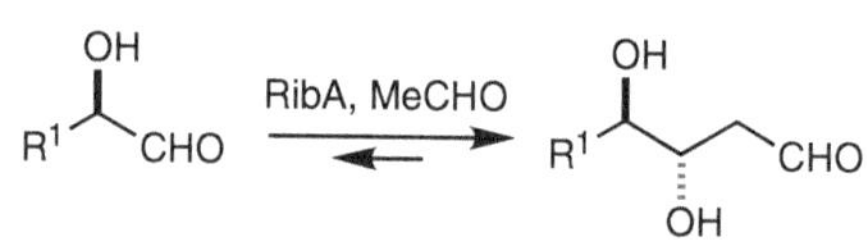

entry	R^1	yield (%)
1	$CH_2OPO_3^{2-}$	78
2	CH_2OH	65
3	Me	32
4	CH_2F	33
5	CH_2Cl	37
6	CH_2Br	30
7	CH_2SH	33

Applications of this transformation were reported in total synthesis of epothilone A. By the use of RibA the two different configured aldehydes **41** and **43** were isolated.[67] Aldehydes **41** and **43** are important key intermediates for construction of fragment A and fragment B in the total synthesis of epothilone A (Scheme 3.6.13).

Chain elongation of the *manno*-configured substrate **44** by NeuA catalysis (NeuA = *N*-acetylneuraminic acid aldolase) yielded the potential intermediate **45** for the total synthesis of amphotericin B with good yields and selectivity (Scheme 3.6.14).[68,69]

The C9–C16 chain fragment **46** of the antibiotic pentamycin was obtained by a FruA-catalysed aldol addition of DHAP with the chiral aldehyde **47** (Scheme 3.6.15, FruA = D-fructose-1,6-bisphosphate aldolase).[70,71]

The metabolism of α-amino-β-hydroxy acids also involves enzymes which catalyse an aldol addition of aldehydes to α-amino acids. These enzymes have

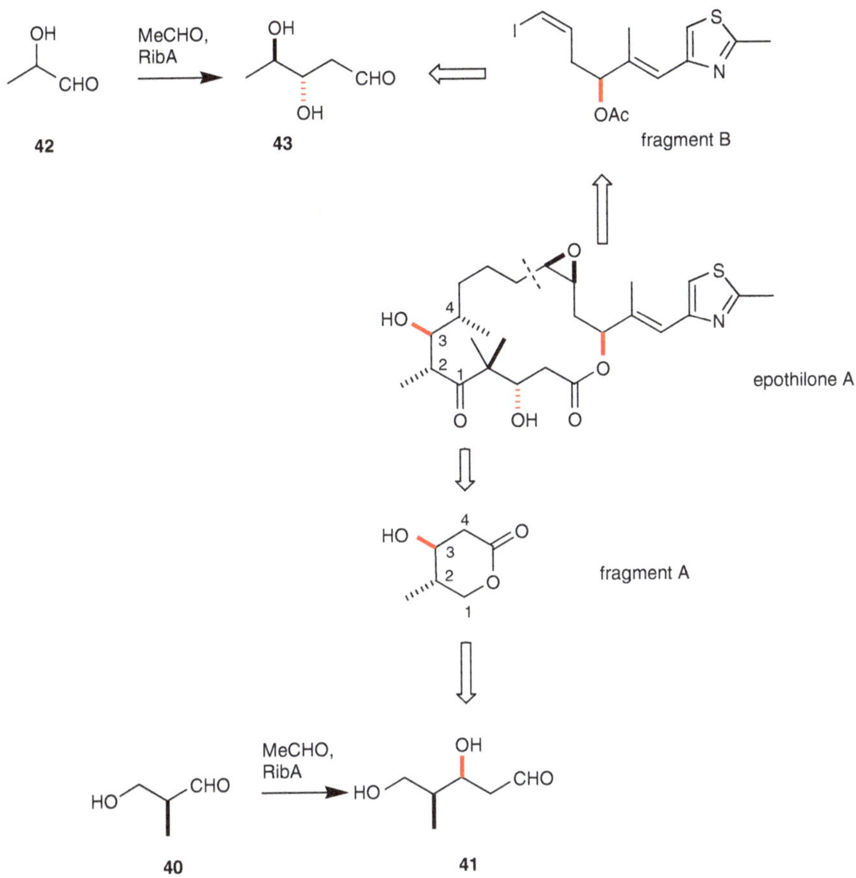

Scheme 3.6.13 Total synthesis of epothilone A using RibA-catalysed aldol additions

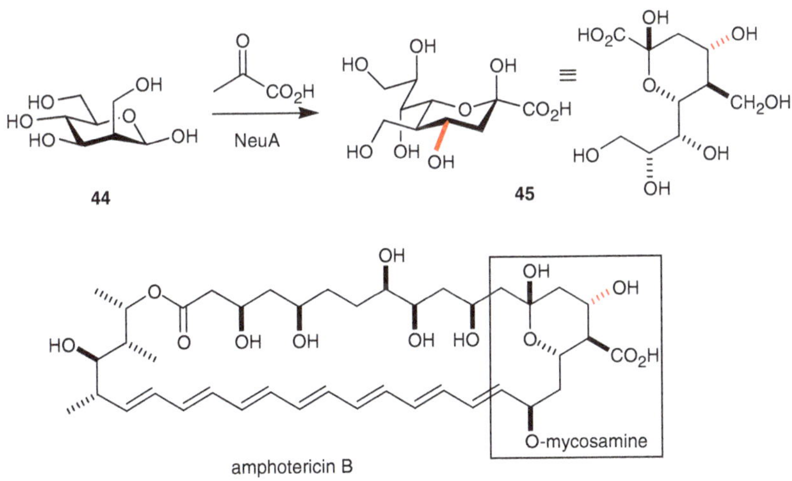

Scheme 3.6.14 NeuA approach to amphotericin B

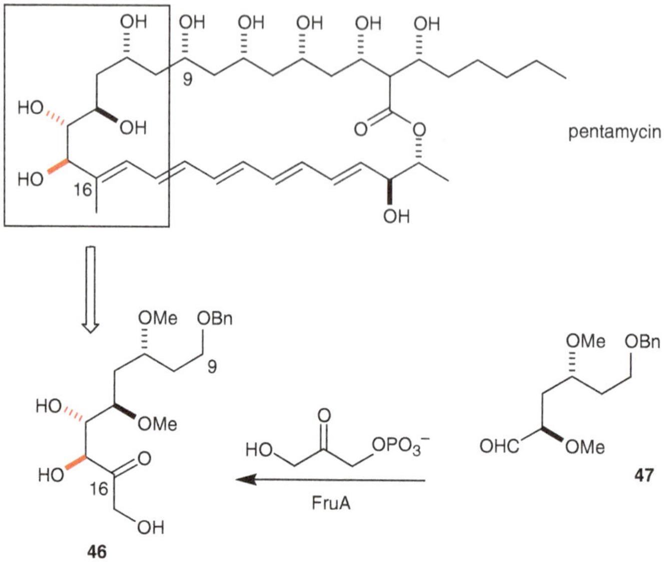

Scheme 3.6.15 FruA-catalysed approach to pentamycin

been used for the synthesis of several chiral α-amino-β-hydroxy acid derivatives and they should be mentioned here.[72,73,74,75] Diastereoselectivity obtained in these transformations are low. Nevertheless, some applications using these enzymes were reported. The synthesis of dihydroxyproline[76] and mycestericin[77] was accomplished in this way.

References

1. Gröger H, Drauz K (2004) In: Asymmetric Catalysis on Industrial Scale, Blaser HU, Schmidt E (eds). Wiley-VCH, Weinheim, p 131
2. Liljeblad A, Kanerva LT (2006) Tetrahedron 62:5831
3. Gijsen HJM, Qiao L, Fitz W, Wong CH (1996) Chem Rev 96:443
4. Fessner WD, Helaine V (2001) Curr Opin Biotechnol 12:574
5. Whalem LJ, Wong CH (2006) Aldrichimica Acta 39:63
6. Tolbert TJ, Wong CH (2004) Encyclopedia of Biological Chemistry 1:307
7. Hanson S, Best M, Bryan MC, Wong CH (2004) Trends in Biochemical Sciences 29:656
8. Fessner WD (200) In: Stereoselective Biocatalysis, Lennarz WJ, Lane MD (eds). Elsevier, Oxford, p 239
9. Schetter B, Mahrwald R (2006) Angew Chem Int Ed 45:7506
10. Ikunaka M (2007) Org Proc Res Dev 11:495
11. (a) Fessner WD (2008) In: Asymmetric Organic Synthesis with Enzymes, Gotor V, Alfonso I, Garcia-Urdiales E (eds). Wiley-VCH, Weinheim, p 275; (b) List B (2007) In: Asymmetric Synthesis, Christmann, Bräse (eds). Wiley-VCH, Weinheim, p 161; (c) Sukumaran J,

Hanefeld U (2005) Chem Soc Rev 34:530; (d) Fessner WD (2004) In: Modern Aldol Reactions, Mahrwald R (ed). Wiley VCH, Weinheim, vol 1, p 201; (e) Silvestri MG, Desantis G, Mitchell M, Wong CH (2003) Top Stereochem 23:267; (f) Roberts SM (2001) J Chem Soc Perkin Trans 1:1475; (g) Hendrix M, Wong CH (1999) Bioorg Chem: Carbohydrates 198; (h) Hendrix M, Wong CH (1999) Bioorg Chem: Carbohydrates 569
12. Kim MJ, Hennen WJ, Sweers HM, Wong CH (1988) J Am Chem Soc 10:6481
13. Sugai T, Kuboki A, Hiramatsu S, Okazaki H, Ohta H (1995) Bull Chem Soc Jpn 68:3581
14. DeSantis G, Zhu Z, Greenberg WA, Wong K, Chaplin J, Hanson SR, Farwell B, Nicholson LWR, Weiner CL (2002) J Am Chem Soc 124:9024
15. Greenberg WA, Varvak A, Hanson SR, Wong K, Huang H, Chen P, Burk MJ (2004) Proc Natl Acad Sci USA 101:5788
16. Reetz MT (2006) Adv Catal 49:1
17. Williams GJ, Woodhall T, Farnsworth LM, Nelson A, Berry A (2006) J Am Chem Soc 128:16238
18. Zhou P, Salleh MH, Honek JF (1993) J Org Chem 58:264
19. Shelton MC, Cotterill IC, Novak STA, Poonawala RM, Sudarshan S, Toone EJ (1996) J Am Chem Soc 118:2117
20. Cotterill IC, Shelton MC, Machemer DEW, Henderson DP, Toone EJ (1998) J Chem Soc Perkin Trans 1:1335
21. Shelton MC, Cotterill IC, Novak STA, Poonawala RM, Sudarshan S, Toone EJ (1996) J Am Chem Soc 118:2117
22. Fessner WD (1992) In: NATO ASI Series, Series C: Mathematical and Physical Sciences, 381:43
23. Takayama S, McGarvey GJ, Wong CH (1997) Ann Rev Microbiol 51:285
24. Fessner WD, Walter C (1996) Top Curr Chem 184:97
25. Toone EJ, Simon ES, Bednarski MD, Whitesides GM (1989) Tetrahedron 45:5365
26. Fessner WD, Eyrisch O (1992) Angew Chem Int Ed Engl 31:56
27. Fessner WD, Walter C (1996) Top Curr Chem 184:97.
28. Fessner WD, Sinerius G, Schneider A, Dreyer M, Schulz GE, Badia J, Aguilar J (1991) Angew Chem Int Ed Engl 30:555
29. Fessner WD, Sinerius G, Schneider A, Dreyer M, Schulz GE, Badia J, Aguilar J. (1991) Angew. Chem. Int. Ed. Engl. 30:555
30. Fessner WD, Badia J, Eyrisch O, Schneider A, Sinerius G (1992) Tetrahedron Lett 33:5231
31. Ziegler T, Straub A, Effenberger F (1988) Angew Chem 100:737
32. Fessner WD (1998) Curr Opin Chem Biol 2:85
33. Duncan R, Drueckhammer DG (1996) J Org Chem 61:438
34. Borysenko CW, Spaltenstein A, Straub JA, Whitesides GM (1989) J Am Chem Soc 111:9275
35. Fessner WD, Gosse C, Jaeschke G, Eyrisch O (2000) Eur J Org Chem 125
36. Wong CH, Alajarin R, Moris-Varas F, Blanco O, Garcia-Junceda E (1995) J Org Chem 60:7360
37. Fessner WD, Sinerius G (1994) Bioorg Med Chem 2:639
38. Chou WC, Chen LH, Fang JM, Wong CH (1994) J Am Chem Soc 116:6191
39. Von der Osten CH, Sinskey AJ, Barbas CF III, Pederson RL, Wang YF, Wong CH (1989) J Am Chem Soc 111:3924
40. Straub A, Effenberger F, Fischer P (1990) J Org Chem 55:3926
41. Pederson RL, Kim MJ, Wong CH (1988) Tetrahedron Lett 29:4645
42. Liu KKC, Kajimoto T, Chen L, Zhong Z, Ichikawa Y, Wong CH (1991) J Org Chem 56:6280
43. Zhou PZ, Salleh HM, Chan PCM, Lajoie G, Honek JF, Nambiar PTC, Ward OP (1993) Carb Res 239:155
44. Lees WJ, Whitesides GM (1992) Bioorg Chem 20:173
45. Romero A, Wong CH (2000) J Org Chem 65:8264
46. Schultz M, Waldmann H, Kunz H, Vogt W (1990) Liebigs Ann 1019

47. Chenevert R, Dasser M (2000) J Org Chem 65:4529
48. Matsumoto K, Shimagaki M, Nakata T, Oishi T (1993) Tetrahedron Lett 34:4935
49. Chenevert R, Lavoie M, Dasser M (1997) Can J Chem 75:68
50. Liu KKC, Kajimoto T, Chen L, Zhong Z, Ichikawa Y, Wong CH (1991) J Org Chem 56:6280
51. Wang YF, Dumas DP, Wong CH (1993) Tetrahedron Lett 34:403
52. Takaoka Y, Kajimoto T, Wong CH (1993) J Org Chem 58:4809
53. Pederson RL, Wong CH (1989) Heterocycles 28:477
54. Castillo JA, Calveras J, Casas J, Mitjans M, Vinardell MP, Parella T, Inoue T, Sprenger GA, Joglar J, Clapes P (2006) Org Lett 8:6067
55. Kajimoto T, Liu KKC, Pederson RL, Zhong Z, Ichikawa Y, Porco JA Jr, Wong CH (1991) J Am Chem Soc 113:6187
56. Henderson I, Laslo K, Wong CH (1994) Tetrahedron Lett 35:359
57. Moris-Varas F, Qian XH, Wong CH (1996) J Am Chem Soc 118:7647
58. Qiao L, Murray BW, Shimazaki M, Schultz J, Wong CH (1996) J Am Chem Soc 118:7653
59. Liu KKC, Kajimoto T, Chen L, Zhong Z, Ichikawa Y, Wong CH (1991) J Org Chem 56:6280
60. Moris-Varas F, Qian XH, Wong CH (1996) J Am Chem Soc 118:7647
61. Liu J, Hsu CC, Wong CH (2004) Tetrahedron Lett 45:2439
62. El Blidi L, Ahbala M, Bolte J, Lemaire M (2006) Tetrahedron: Asymm 17:2684
63. Eyrisch O, Fessner WD (1995) Angew Chem Int Ed Engl 34:1639
64. Chen L, Dumas DP, Wong CH (1992) J Am Chem Soc 114:741
65. Gijsen HJM, Wong CH (1995) J Am Chem Soc 117:2947
66. Gijsen HJM, Wong CH (1994) J Am Chem Soc 116:8422
67. Liu J, Wong CH (2002) Angew Chem Int Ed 41:1404
68. Koppert K, Brossmer R (1992) Tetrahedron Lett 33:8031
69. Malleron A, David S (1996) New J Chem 20:153
70. Matsumoto K, Shimagaki M, Nakata T, Oishi T (1993) Tetrahedron Lett 34:4935
71. Shimagaki M, Muineshima H, Kubota M (1993) Chem Pharm Bull 41:282
72. Kimura T, Vassilev VP, Shen GJ, Wong CH (1997) J Am Chem Soc 119:11734
73. Shibata K, Shingu K, Vassilev VP, Nishide K, Fujita T, Node M, Kajimoto T, Wong CH (1996) Tetrahedron Lett 37:2791
74. Saeed A, Young DW (1992) Tetrahedron 48:2507
75. Steinreiber J, Fesko K, Reisinger C, Schuermann M, van Assema F, Wolberg M, Mink D, Griengl H (2007) Tetrahedron 63:918
76. Fujii M, Miura T, Kajimoto T, Ida Y (2000) Synlett 1046
77. Shibata K, Shingu K, Vassilev VP, Nishide K, Fujita T, Node M, Kajimoto T, Wong C-H (1996) Tetrahedron Lett 37:2791

3.7 Organocatalysed Aldol Addition

In the field of catalytic and enantioselective aldol additions the area of organocatalysed aldol addition has shown the highest rates of increase over the last 10 years. Hence the organocatalytic methods are developed at a very rapid pace. Numerous reports were published and for that reason only a selection of the most important results can be given here. Aldol additions and condensations of aldehydes and ketones in the presence of amines have been known for a long time. But the full potential of their synthetic utility especially with regard to stereoselective and catalytic execution has been discovered systematically over the last 10 years only. For comprehensive overviews of this development see reference.[1,2,3]

The prototype of this transformation is the working mode of class I aldolases. These enzymes bind the substrate temporary and covalently via an enamine-imine formation as it is shown in Scheme 3.7.1. A lysine residue can then initiate C–C bond formation or cleavage.

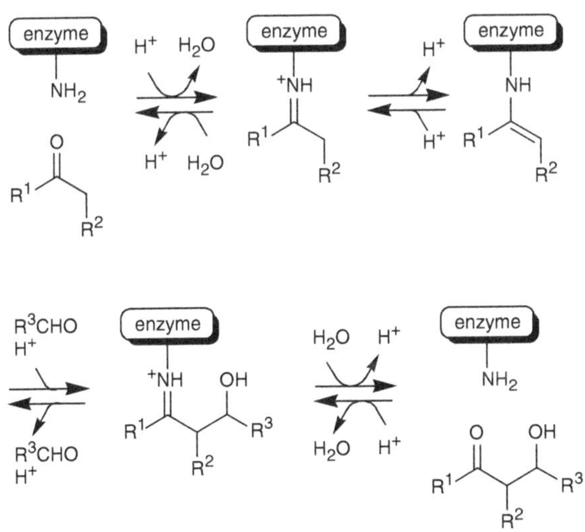

Scheme 3.7.1 Reaction mechanism of enzymes as the prototype for amine-catalysed aldol reactions

In contrast to this knowledge and mechanistic investigations[4,5,6,7] only a few reports on preparative utilization of this mode of an intermolecular aldol reaction have appeared in the literature before the end of the last century.[8,9,10,11,12] The reactions were performed in the presence of primary or secondary amines in combination with a carboxylic acid. An early example of a diethylamine- and piperidine-catalysed aldol process is given in Scheme 3.7.2.[13,14]

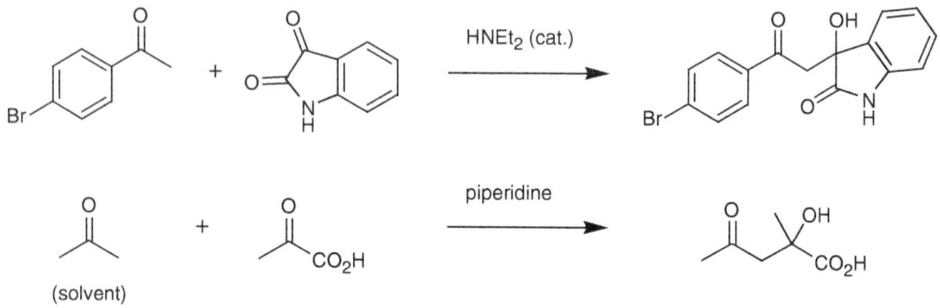

Scheme 3.7.2 Diethylamine- and piperidine-catalysed aldol additions

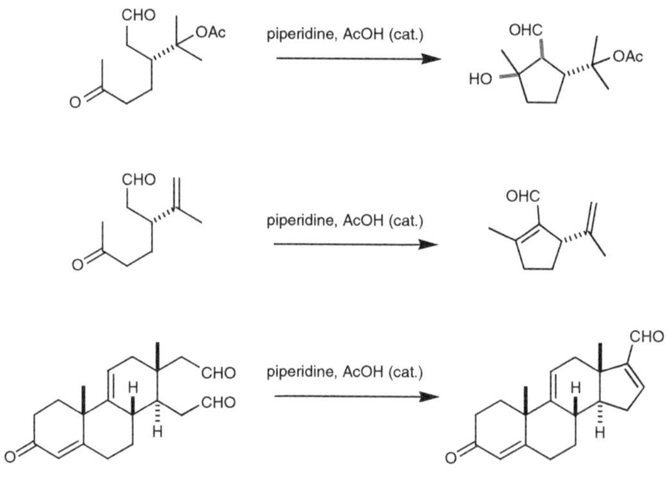

1 2

Scheme 3.7.3 Intramolecular piperidine-catalysed aldol additions

Since the reaction conditions were often harsh, dehydrations occurred and the corresponding α,β-unsaturated ketones were obtained. On the other hand – especially in total synthesis of steroids – β-hydroxyketones are often intermediates when the double bond formation is desired in cyclic systems.[15] Diethylamine and piperidine catalysts were also successfully used in the Knoevenagel transformations.[16] The more explored field in the beginning of this development represents intramolecular aldol reactions. They are divided into enolendo- and enolexo-cyclization. This classification depends on the formation of enols during the reaction (kinetic or thermodynamic controlled enol). One of the first successful applications was reported by Woodward and coworkers. During the total synthesis of steroid **2** the D-ring was constructed by an enolexo aldol condensation of the 1,6-dialdehyde **1** in the presence of catalytic amounts of piperidine and acetic acid (Scheme 3.7.3).[17] Further examples of 5-enolexo aldol reactions are given in Scheme 3.7.3.[18,19,20,21,22,23]

A very instructive example of cyclization depending on the reaction conditions is shown in Scheme 3.7.4. By using different reaction conditions – amine or inorganic base catalysis – different enolizations were obtained. Thus, an approach to different substituted cyclopentenes is given.[24]

Scheme 3.7.4 Regioselective 5-enolexo aldol reaction with respect to the reaction conditions

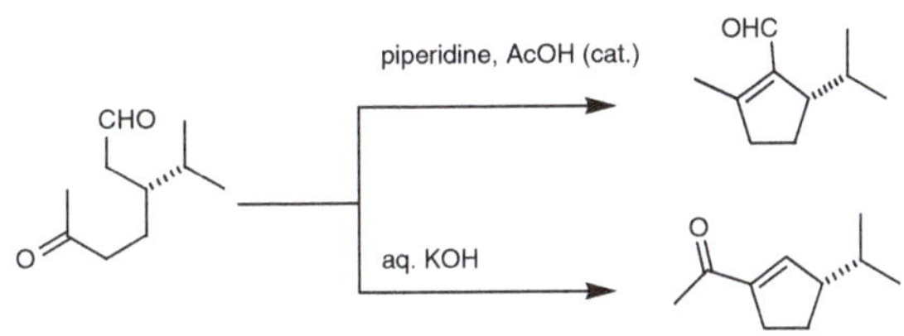

A selection of further examples of 6-enolexo and 6-enolendo aldol reactions is listed in Scheme 3.7.5.[25,26,27,28,29,30]

A very important industrial application of amine-catalysed intramolecular cyclization represents the construction of bicyclic ketones – a version of Robinson annulation (Scheme 3.7.6). This cyclization is achieved via pyrrolidine catalysis and represents an easy approach to very important intermediates in the total synthesis of steroids.[31,32]

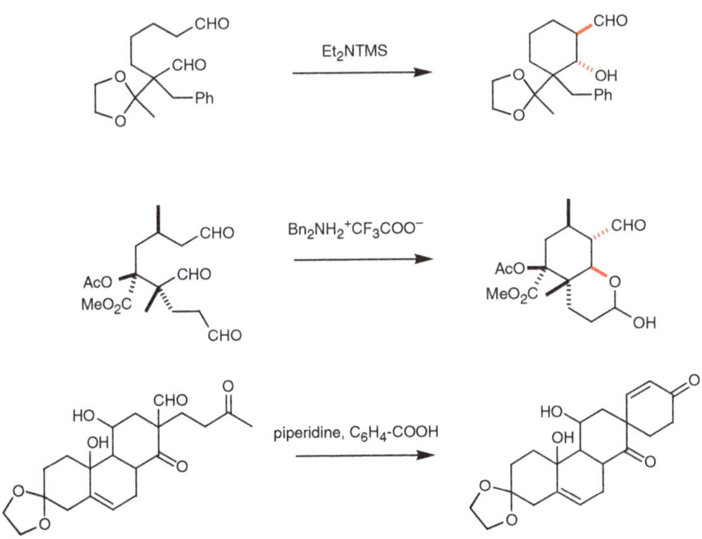

Scheme 3.7.5 Examples for 6-enolexo and 6-enolendo cyclizations

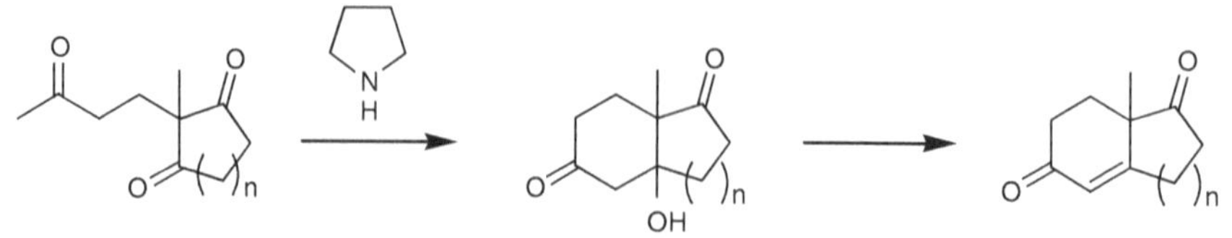

Scheme 3.7.6 Pyrrolidine-catalysed intramolecular aldol condensation

An enantioselective approach of this cyclization was first reported by Eder, Wiechert and Sauer in 1971.[33] They used catalytic amounts of (*S*)-proline in the presence of perchloric acid to obtain the bicyclic aldol condensation products **4** and **6** with a high degree of enantioselectivity. Hajos and Parrish at Hoffmann La Roche independently developed a similar process.[34] They obtained the aldol adduct with comparable enantioselectivities as described by Eder, Sauer and Wiechert. Subsequently, dehydratization yielded the condensation products **4** and **6** (Scheme 3.7.7).

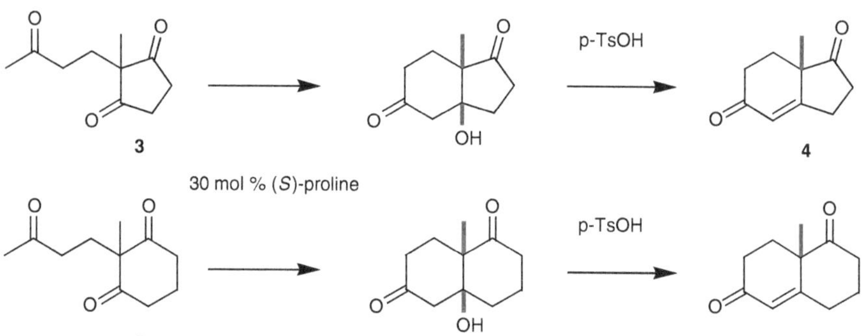

Scheme 3.7.7 Proline-catalysed enantioselective Hajos-Wiechert aldol reaction

For an exceptional behaviour (5-enolexo-cyclization) during an organocatalysed cyclization to Wieland–Miescher ketone-like bicyclic compounds see Hayashi et al.[35]

As pointed out, the full potential of this transformation was rediscovered 20–30 years later.[36] Meanwhile several other amino acids were tested in these reactions[37,38,39], but proline turned out to be the catalyst of choice. Moreover, it seems that this methodology is the most thoroughly theoretically investigated one if one compares the output of publication to that of other methods. Several different transition state models were proposed (Hajos model,[40] Agami model,[41,42] Swaminathan model,[43] Houk model.[44,45]) Several comprehensive overviews of the proline-catalysed aldol addition have been published recently.[46,47,48,49,50,51,52,53] For all these reasons only selected and prominent examples can be discussed here.

Scheme 3.7.8 represents a few examples of proline-catalysed intramolecular aldol additions.[54] Synthesis of **11** represents an 5-enolexo aldol addition. The stereoselectivity observed is lower as in the corresponding 6-enolexo-examples **7–10**.

This is true for both diastereoselectivity and enantioselectivity. An application of the proline-catalysed 6-enolexo-aldolization was reported in the total synthesis of (+)-cocaine.[55]

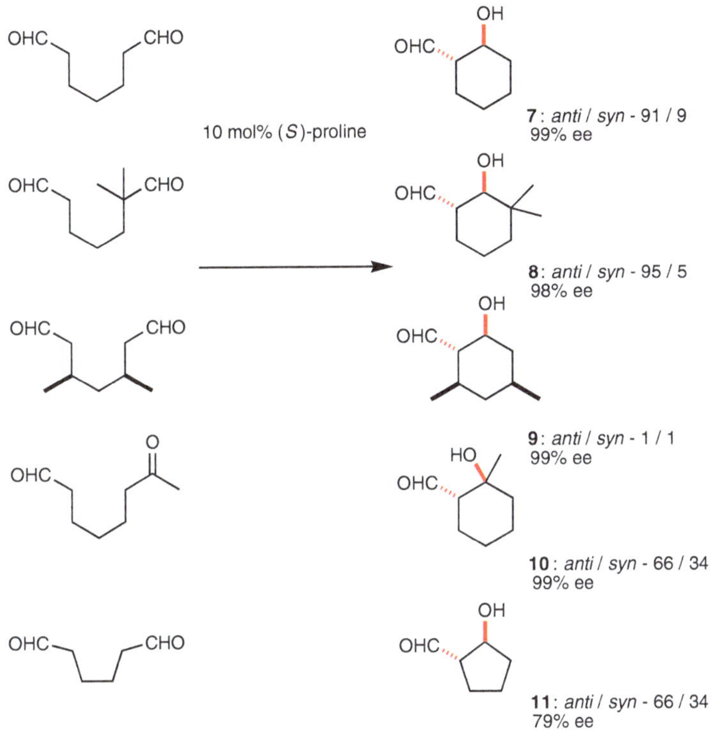

Scheme 3.7.8 Examples for intramolecular aldol reactions catalysed by (*S*)-proline

Recently, examples of a transannular enantioselective aldol addition have been reported. Cyclic diketones can be transformed into the corresponding optically active β-hydroxy ketones by the use of fluoro-substituted proline[56] (Scheme 3.7.9). This method was applied in the total synthesis of hirsutene.

A desired goal of aldol additions in total synthesis of polyketides is the straightforward intermolecular reactions. The enantioselective execution of this transformation poses an important tool to install desired and defined configuration during the construction of carbon skeletons in natural product synthesis. A breakthrough in this context was the first enantioselective, intermolecular, proline-catalysed aldol addition reported in 2000.[57] The authors have demonstrated enantioselective proline-catalysed aldol additions of acetone with a variety of aldehydes. The aldol adducts were obtained with a high degree of enantioselectivity, even in the enolizable series (Scheme 3.7.10). α-Unbranched aldehydes reacted in these transformations with lower yields. In addition, lower enantioselectivities were detected in the aldol adducts. Much more side reactions were observed. Acetone was used in great excess to keep side reactions at a minimum.

Scheme 3.7.9 Transannular enantioselective proline-catalysed aldol addition

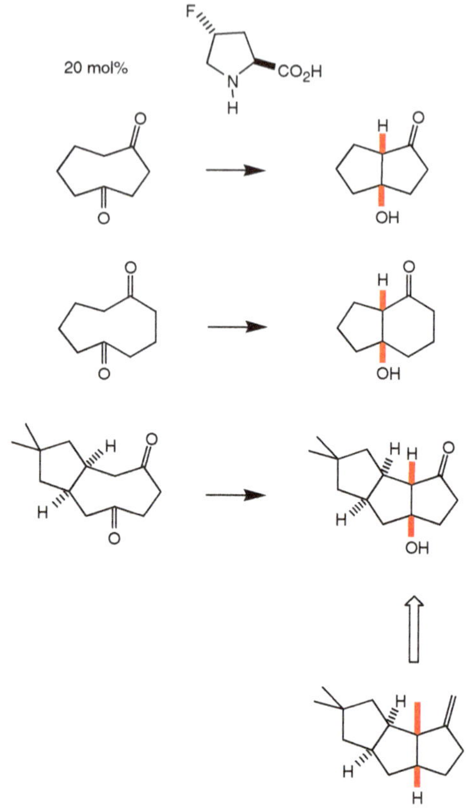

12: 53%
94% ee

13: 57%
82% ee

14: 84%
96% ee

(+)–hirsutene

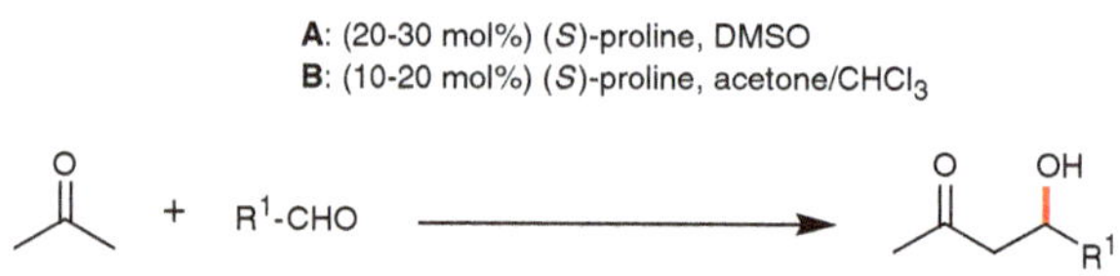

entry	R^1	conditions	yield (%)	ee (%)
1	4-NO_2-C_6H_4	A	68	76
2	Ph	A	62	72
3	*i*-Pr	A	97	96
4	*t*-Bu	A	81	>99
5	Cy	A	63	84
6	H_2C=$CHCH_2C(CH_3)_2$	A	85	>99
7	*n*-Bu	B	31	67
8	H_2C=$CHCH_2CH_2$	B	34	72
9	*i*-$PrCH_2$	B	24	73

Scheme 3.7.10 Intermolecular enantioselective aldol reactions of aldehydes and acetone catalysed by proline

Substituted acetone derivatives were also applied to the proline-catalysed aldol addition. The aldol adducts were isolated with a high degree of enantioselectivity, but with moderate diastereoselectivity, in some examples even without any diastereoselectivity (Scheme 3.7.11).[58,59] For employment of alanine in these aldol additions see Cordova et al.[60] and for application of tryptophane see Jiang et al.[61]

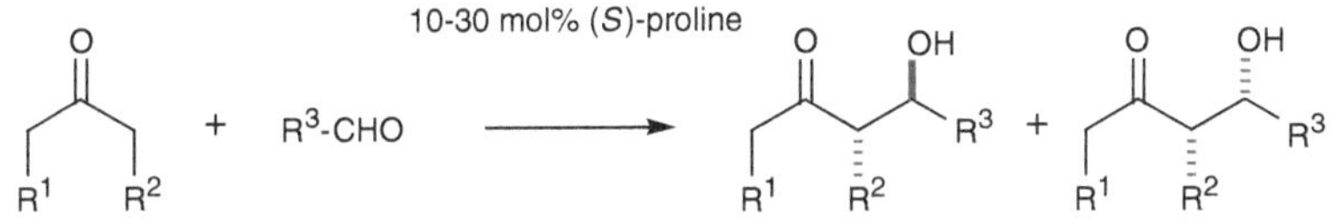

entry	R^1	R^2	R^3	yield (%)	*anti* / *syn*	ee (%,*anti*)	ee (%,*syn*)
1	$CH_2CH_2CH_2$		Ph	85	50 / 50	85	76
2	$CH_2CH_2CH_2$		*i*-Bu	41	88 / 12	86	89
3	$CH_2CH_2CH_2$		*i*-Pr	68	>95 / 5	97	-
4	CH_2CH_2		*i*-Bu	77	75 / 25	95	20
5	Me	OH	*i*-Pr	62	>95 / 5	>99	-
6	Me	OH	*t*-Bu CH_2	38	63 / 37	97	84
7	Me	OH	2-Cl-C_6H_4	95	60 / 40	67	32

Scheme 3.7.11 Proline-catalysed asymmetric aldol reaction to α-substituted acetone

Also, α-branched aldehydes were applied to the chiral amine-catalysed aldol addition. Different proline-derived chiral amino alcohols and diamines were tested as catalysts in these reactions. Thus, an organocatalytic approach to stereogenic quaternary carbon centres was elaborated. Moderate diastereoselectivities were detected in the isolated aldol adducts. The *anti*-configured aldol adducts were obtained with high degrees of enantioselectivity. Some selected results are shown in Scheme 3.7.12.[62] For a racemic version of this process in the presence of pyrrolidine see also Mase et al.[63]

The enantioselective cross-aldol addition of two different aldehydes represents a very promising tool for total synthesis of natural products. MacMillan and coworkers reported the first successful execution of this reaction. Following this protocol they were able to isolate *anti*-configured β-hydroxyaldehydes with a high degree of both enantioselectivity and diastereoselectivity (Scheme 3.7.13).[64]

A proline-catalysed cross-aldol addition in the total synthesis of belactosin C is described in Kumaraswamy and Markondaiah.[65]

In the following time a vast amount of reports were published dealing with the modification of catalysts in order to improve the yields, reaction rates and enantioselectivities.[66] Even a DNA-tethered proline was tested in aldol additions of acetone and aromatic aldehydes.[67] For the first report using DNA directly as an organocatalyst in aldol additions see Sun et al.[68] An initial and first DNA-templated

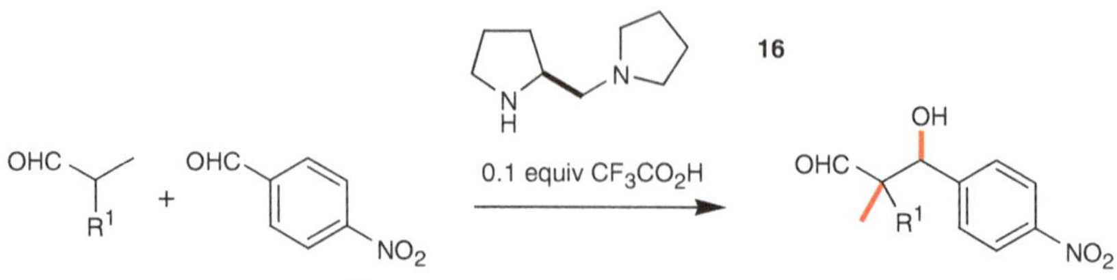

entry	R^1	yield (%)	*anti* / *syn*	ee (%,*anti*)	ee (%,*syn*)
1	Et	96	62 / 38	91	75
2	*i*-Pr	92	66 / 34	89	66
3	$H_3C(CH_2)_8$	93	69 / 31	91	68
4	$Me_2C{=}CHCH_2CH_2$	96	65 / 35	89	52
5	4-*i*-$PrC_6H_4CH_2$	97	84 / 16	95	74
6	4-*t*-$BuC_6H_4CH_2$	91	85 / 15	96	68

Scheme 3.7.12 Aldol reactions between enolizable aldehydes and 4-nitrobenzaldehyde catalysed by chiral diamine **16**

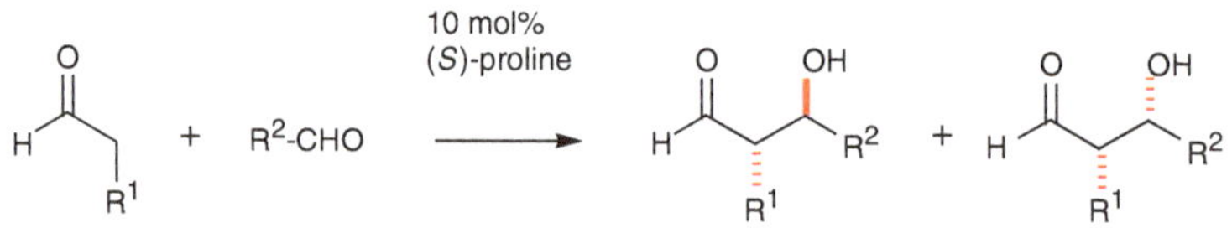

entry	R^1	R^2	yield (%)	*anti* / *syn*	ee (%,*anti*)
1	Me	Et	80	80 / 20	99
2	Me	CH_2*i*-Pr	88	75 / 25	97
3	Me	Cy	87	93 / 7	99
4	Me	Ph	81	75 / 25	99
5	Me	*i*-Pr	82	96 / 4	>99
6	Bu	*i*-Pr	80	96 / 4	98
7	Bn	*i*-Pr	75	95 / 5	91

Scheme 3.7.13 Enantioselective proline-catalysed cross-aldol reaction between enolizable aldehydes

cross-aldol reaction of glyceraldehyde and glycolaldehyde could be successfully executed in the presence of various amine-containing catalysts. Thus, an organocatalysed approach to linked riboses in DNA could be realized.[69]

For the use of polymer-supported proline[70] and proline-catalysed aldol additions in aqueous media see Darbre and Machuqueiro and others.[71] Recent investigations have been characterized by the application of this catalytic transformation to different substituted substrates.[72,73,74]

Typical examples for proline derivative-catalysed aldol additions in water are depicted in Scheme 3.7.14.[75]

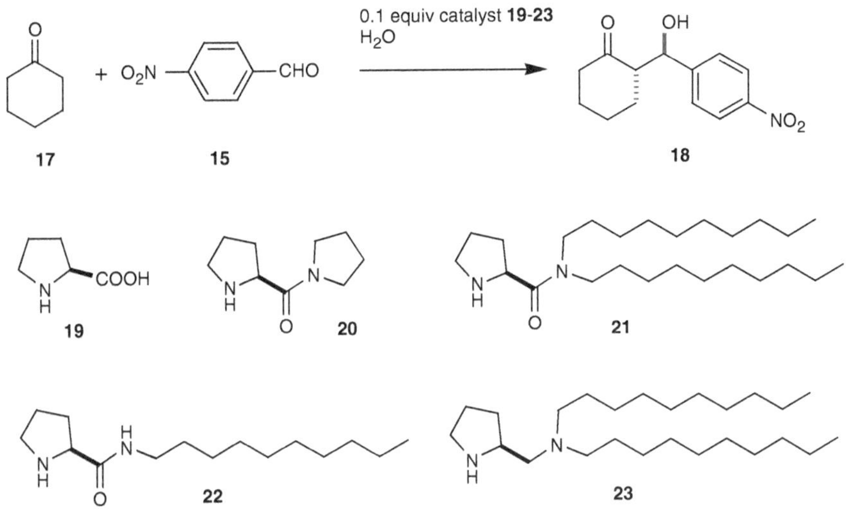

entry	catalyst	additive	yield (%)	*anti* / *syn*	ee (%, *anti*)
1	**19**[a]	-	65	63 / 37	89
2	**19**	-	0	-	-
3	**20**	-	68	84 / 16	3
4	**21**	-	78	84 / 16	22
5	**22**	-	99	79 / 21	36
6	**23**	(+)–CSA	99	68 / 32	85
7	**23**	$Sc(OTf)_3$	94	84 / 16	93
8	**23**	TFA	99	89 / 11	94

Scheme 3.7.14 Proline-derivative-catalysed aldol addition in water;[a] in DMSO

An example for the application of the proline-catalysed aldol addition is given in total synthesis of epothilone. A straightforward approach via proline catalysis to a chiral key product was described by Avery and Zheng. They isolated the chiral hydroxydiketone **25** in ee's over 99% by an (*R*)-proline-catalysed aldol addition of acetone and ketoaldehyde **24** (Scheme 3.7.15).[76] For comparative studies in the lithium enolate series of this aldol addition see Nicolaou et al.[77]

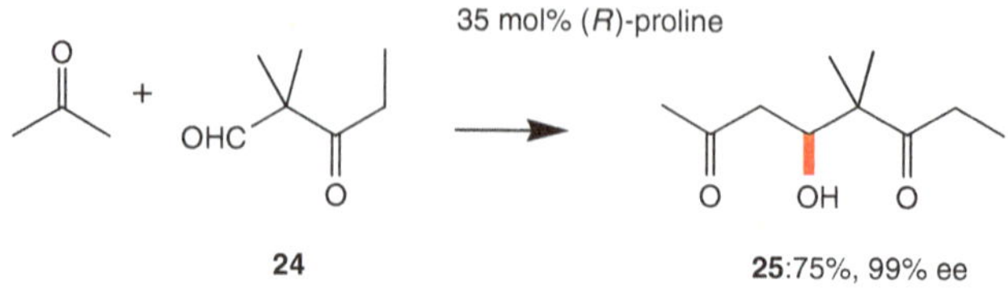

Scheme 3.7.15 (*R*)-Proline-catalysed synthesis of intermediates of epothilone

A very similar key building product was synthesized by an (*S*)-proline-catalysed aldol reaction (Scheme 3.7.16). The aldol adduct was used in the total synthesis of ipsenol.[78]

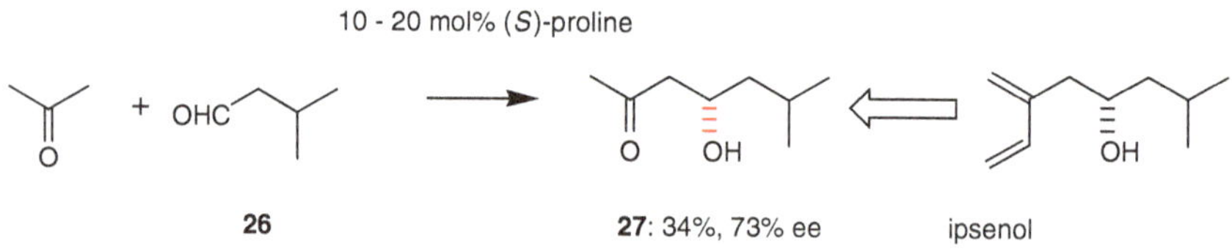

Scheme 3.7.16 (*S*)-Proline-catalysed total synthesis of ipsenol

Optically pure aldol adduct of pivaldehyde and acetone was used as the starting material in the total synthesis of apratoxin A.[79] The required (*S*)-configuration was installed by an (*R*)-proline-catalysed aldol addition.[80] Several examples of proline-catalysed aldol reactions in total synthesis of natural products are shown in the following schemes. The chiral centre of trichostatin A was created by an (*S*)-proline-catalysed aldol addition. The *anti*-configured aldol product **28** was obtained with high degrees of diastereoselectivity as well as enantioselectivity (Scheme 3.7.17).[81]

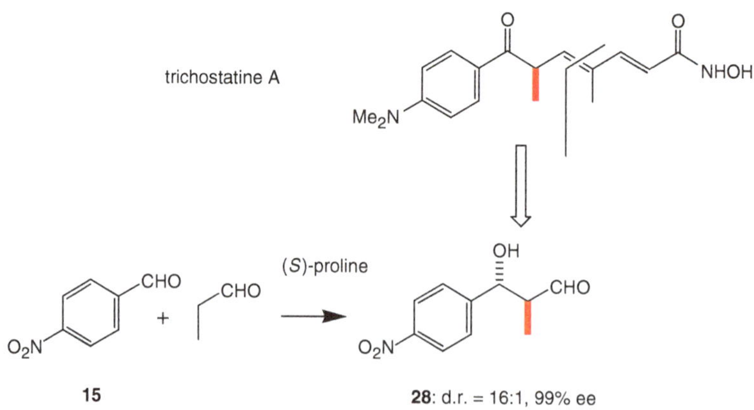

Scheme 3.7.17 (*S*)-Proline-catalysed aldol addition in total synthesis of trichostatin A

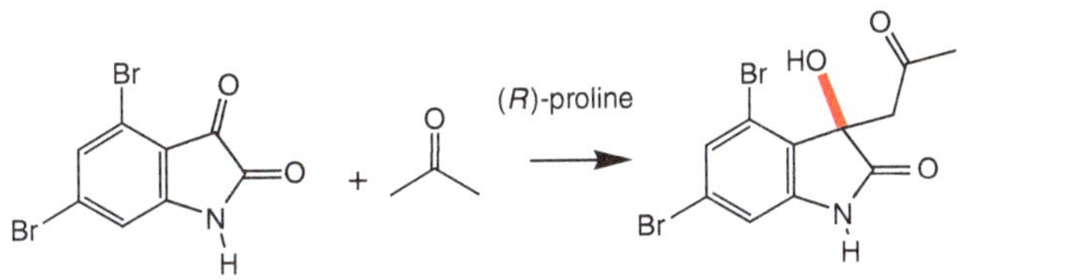

29

30: 55% ee
97% ee after recrystallization

Scheme 3.7.18 Total synthesis of (*R*)-convolutamydine A

Proline-catalysed aldol addition of ketones to ketones was described during the total synthesis of (*R*)-convolutamydine A. By the aldol addition of acetone to dibromoisatine the tertiary alcohol **30** is obtained with only low enantioselectivities (55% ee). But by crystallization nearly optically pure aldol adduct can be obtained.[82]

An asymmetric intramolecular proline-catalysed aldol reaction was realized during the total synthesis of (+)-juvabione. By the use of 25 mol% of silylated hydroxyproline **32** the bicyclic aldol adduct **33** was obtained with high degrees of enantioselectivity. Again by crystallization optically pure aldol adduct can be obtained.[83] The same aldol adduct **33** was used in the total synthesis of a cannabinoid receptor agonist CP 55, 940 (Schemes 3.7.19).[84]

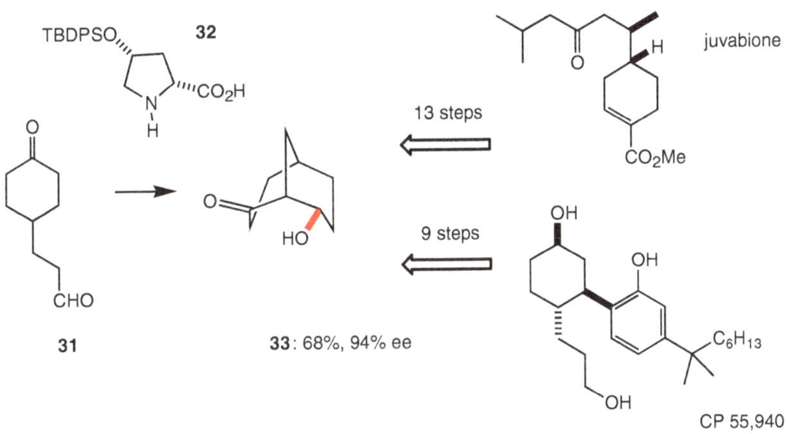

Scheme 3.7.19 Total synthesis of juvabione and CP 55,940

Also, cyclic ketones were used in proline-catalysed aldol reactions (Scheme 3.7.20). By the use of (*S*)-proline in aldol additions of cyclopentanone and aldehyde **34** an access to optically active acetoxyhexadecanolide – an oviposition attractant pheromone – is given.[85] Phytosphingosines are accessible by the use of the Enders-dioxanone in (*S*)-proline-catalysed aldol addition.[86] Both aldol adducts **35** and **37** were isolated with a high degree of enantioselectivity (Scheme 3.7.20). For a similar application of (*R*)-proline in total synthesis of jaspine B see Enders et al.[87]

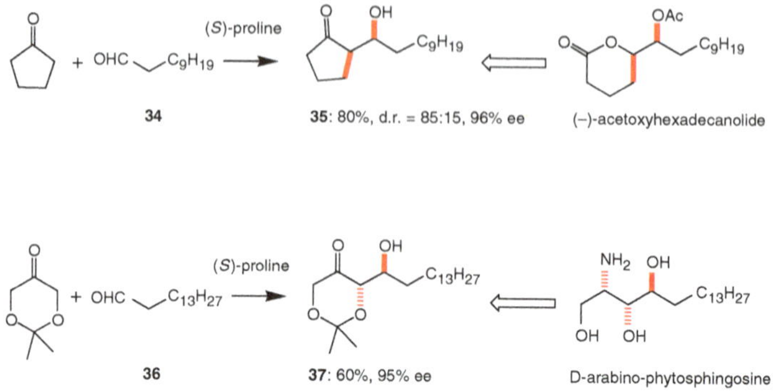

Scheme 3.7.20 Proline-catalysed aldol reactions in total synthesis of (–)–acetoxyhexadecanolide and phytosphingosines

For a proline-catalysed access to phytosphingosines by aldol additions of α-amino aldehydes with cyclic ketones see Kumar and Rode.[88]

As described, unfunctionalized and mostly aromatic aldehydes have been employed in the proline-catalysed aldol reaction so far. The more challenging task turns out to be organocatalysed aldol additions of oxygen-funtionalized aldehydes or ene-components. By this way an access to configurative defined hydroxylated polyketides is possible. Only a few examples of this strategy have been published so far.

In 2000, List and Notz described the first proline-catalysed enantioselective aldol addition of unprotected hydroxyacetone **38** with several enolizable aliphatic aldehydes.[58] High regioselectivities (>20:1) and extremely high enantioselectivities (>100:1) were detected. The diastereoselectivities observed depended on the aldehydes used in these reactions. Even protected glyceraldehyde was reacted with hydroxyacetone. Only moderate 1,2-asymmetric induction was observed during this transformation. Fructose and tagatose derivatives **39f** were isolated with only moderate diastereoselectivities (entry 6, Scheme 3.7.21).

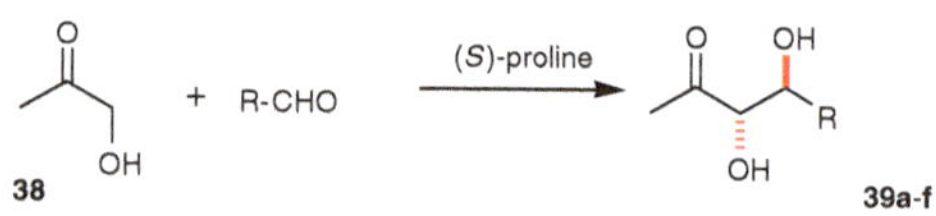

entry	R	yield (%)	*anti* / *syn*	ee (%)
1	Cy	60	>95 / 5	>99
2	*i*-Pr	62	>95 / 5	>99
3	PhCH(Me)-	51	>95 / 5	>95
4	2-Cl-C_6H_4-	95	60 / 40	67
5	*t*-Bu-CH_2-	38	63 / 37	>97
6	glyceraldehyde	40	67 / 33	>97

Scheme 3.7.21 Proline-catalysed aldol additions of hydroxyacetone

Shortly after that, Barbas and coworkers reported the first organocatalysed aldol addition of unprotected dihydroxyacetone (DHA) with acetonide protected glyceraldehyde. The reaction was catalysed by chiral diamines derived from proline in an aqueous phosphate buffer.[89] No 1,2-asymmetric induction was observed in this reactions. Protected D-fructose (one of four possible sugars was formed in this reaction) was obtained under these reaction conditions. Recently, several groups have reported organocatalysed aldol additions of aldehydes to hydroxyacetone[90] or derivatives of dihydroxyacetone.[91] The main results of this tremendous work in the DHA-series are summarized in Scheme 3.7.22.

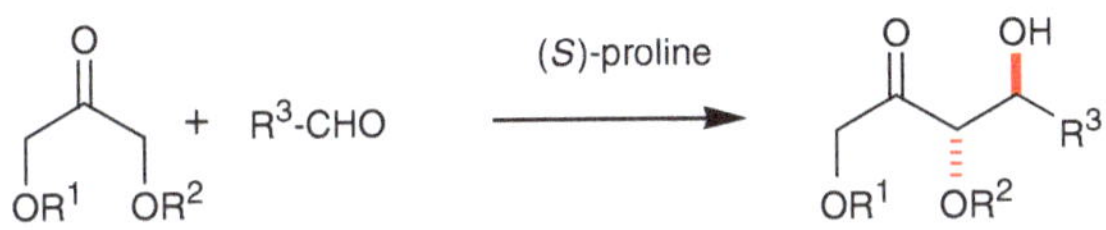

40a-e

entry	R^1	R^2	R^3	yield (%)	*anti* / *syn*	ee (%)
1	H	H	p-NO_2-C_6H_4	-	-	-
2	Bn	Bn	p-NO_2-C_6H_4	-	-	-
3	TIPS	TIPS	p-NO_2-C_6H_4	-	-	-
4	H	TMS	p-NO_2-C_6H_4	-	-	-
5	H	Bn	p-NO_2-C_6H_4	-	-	-
6	-C(Me_2)$_2$-		p-NO_2-C_6H_4	90	6 / 1	96
7	-C(C_5H_{10})-		p-NO_2-C_6H_4	62	5 / 1	67
8	-C(Me_2)$_2$-		$AcOCH_2$	60	>15 /1	98
9	- CH_2 -		p-NO_2-C_6H_4	91	15 / 1	94
10	-C(Me_2)$_2$-		glyceraldehyde	40	n.r.	n.r.

Scheme 3.7.22 Proline-catalysed aldol additions with protected dihydroxyacetone

The results of Scheme 3.7.22 clearly demonstrate that unprotected DHA is not a useful ene-component for the proline-catalysed aldol addition. Furthermore, several other protecting groups are also not suitable for this transformation (entries 2–5, Scheme 3.7.22). Concerning diastereoselection, mainly *anti*-configured up to non-selective aldol adducts were obtained.

Scheme 3.7.23 summarizes further investigations reported by Enders and coworkers. These results were obtained in aldol additions with protected derivatives of glyceraldehyde as well as Garner aldehyde.[92] When used with α-chiral aldehydes in the presence of optional use of (*R*)- or (*S*)-proline matched/mismatched situations become apparent (compare **42d** with **42e** of Scheme 3.7.23 and **42c** of Scheme 3.7.23 with entry 10 in Scheme 3.7.22).

Scheme 3.7.23 (*S*)-Proline-catalysed aldol additions of cyclic acetals of dihydroxyacetone; (*R*)-proline was used as catalyst for compounds **42c** and **42d**

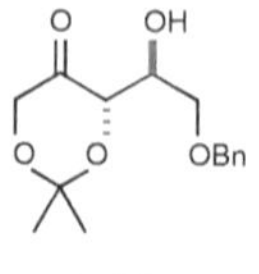

42a:40%
d.r.: >98:2
97% ee

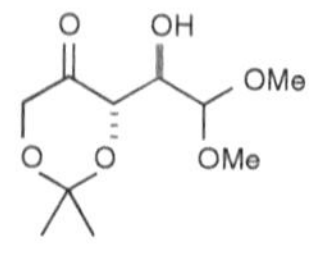

42b:69%
d.r. = 94:6
93% ee

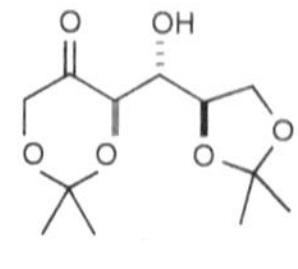

42c:76%
d.r. >98:2
98% ee

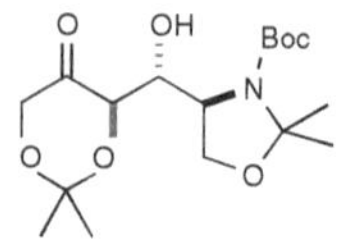

42d: 31%
d.r. >98:2
>96% ee

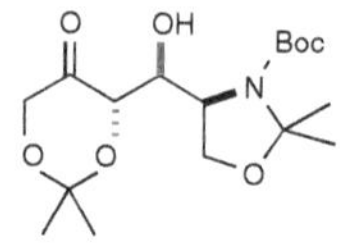

42e: 80%
d.r. >98:2
>96% ee

Later on, Barbas and coworkers described a second type of organocatalysed aldol addition of aldehydes to unprotected DHA **43**. These reactions were carried out in the presence of catalytic amounts of tryptophane or threonine derivatives. By this protocol the aldol adducts of aromatic aldehydes were isolated with a high degree of *syn*-diastereoselectivity as well as enantioselectivity (Scheme 3.7.24).[93]

Scheme 3.7.24 Threonine-catalysed aldol additions of unprotected dihydroxyacetone

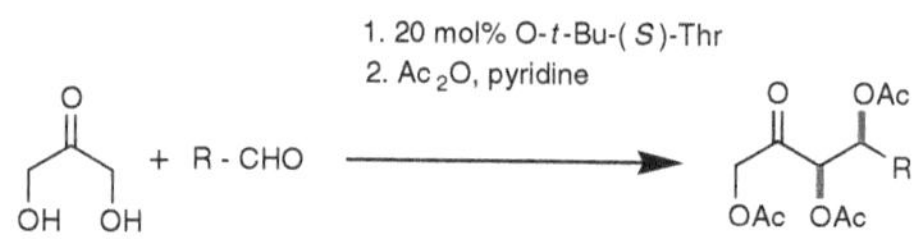

43

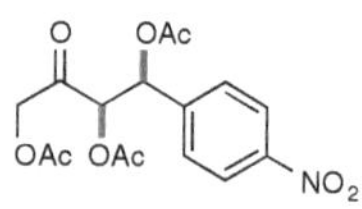

44a: 76%, d.r. = 94:6,
ee (*syn*/*anti*) - 92 / 20

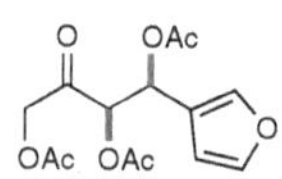

44b: 72%, d.r. = 88:12,
ee (*syn*/*anti*) - 92 / 62

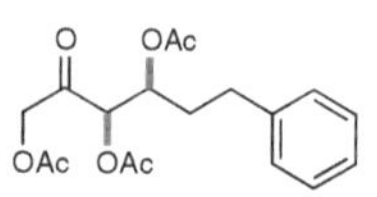

44c : 21%, d.r. = 83:17,
ee (*syn*/*anti*) - 99 / 28

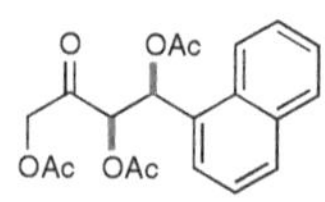

44d: 65%, d.r. = 92/8,
ee (*syn*/*anti*) - 97 / n.d.

Also very recently, Barbas and coworkers demonstrated the utility of threonine and tryptophane derivatives in asymmetric organocatalysed aldol additions with protected DHA **45**.[94] Under these conditions the authors were able to isolate aldol adducts of TBS-protected DHA and enolizable aldehydes with high degrees of enantioselectivity and with good *syn*-diastereoselectivity (**46a–e**, Scheme 3.7.25).

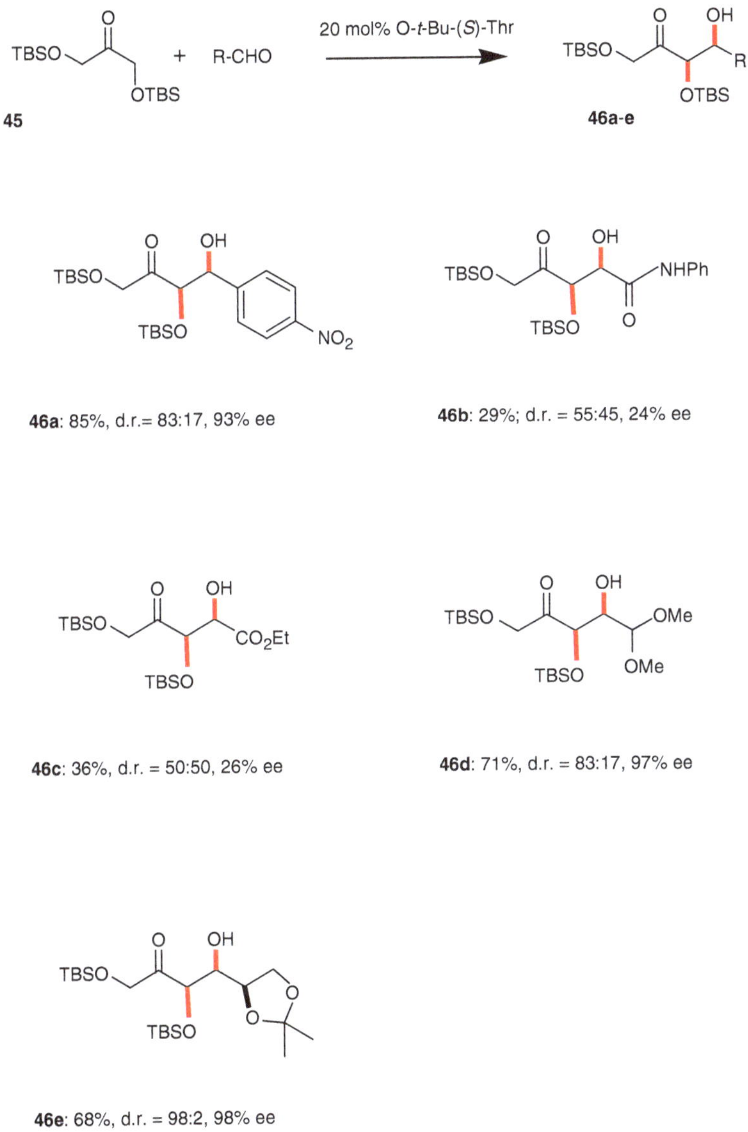

Scheme 3.7.25 Threonine-catalysed aldol additions of protected dihydroxyacetone; O-*t*Bu-(*R*)-threonine was used as catalyst for compound **46e.**

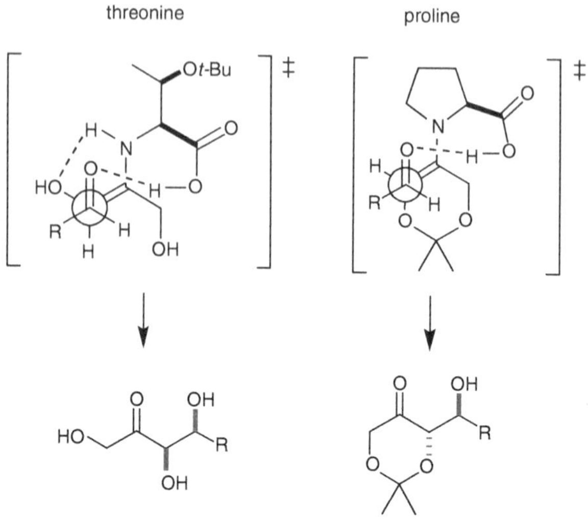

Fig. 3.7.1 Proposed transitions states of proline- and threonine-catalysed aldol additions

For transition states explaining the *anti*-configuration by application of proline as well as *syn*-configuration in the thereonine series see Figure 3.7.1.

α-Hydroxyketones react with aldehydes in the presence of tertiary amines without any additives. The expected aldol adducts were isolated with a high degree of *syn*-diastereoselectivity. The choice of tertiary amine is crucial and depends on the substrates used. Best results so far were obtained in reactions of hydroxyacetone **38** by the application of 5 mol% of DBU. Moreover, an extremely high regioselectivity was observed. The C–C bond formation process took place only at the oxygen-containing α-side of hydroxyketone (Scheme 3.7.26).[95]

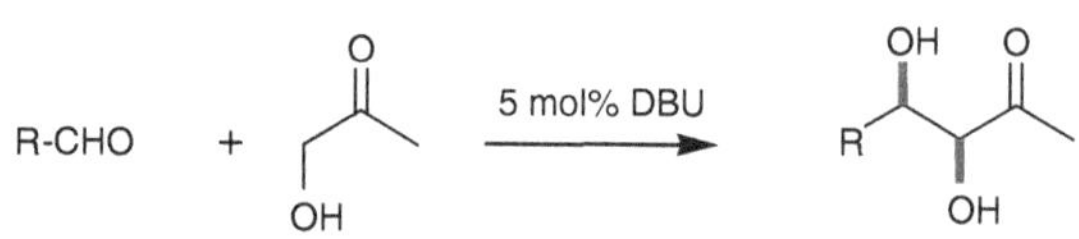

entry	R	yield (%)	*syn* / *anti*
1	*i*-Pr	83	91 / 9
2	$PhCH_2CH_2$–	87	77 / 23
3	Cy	92	90 / 10
4	Ph	89	68 / 32

Scheme 3.7.26 Amine-catalysed aldol additions to hydroxyacetone **38**

In the dihydroxyacetone series Hünig base was the tertiary amine of choice. The *syn*-diastereoselectivity was extremely high – no *anti*-configured aldol products could be obtained (Scheme 3.7.27). When used with hydroxyacetone **38** the corresponding 1,2-diolketones were isolated, whereas by deployment of DHA **43** the corresponding hemiketals of the aldol adducts were obtained.

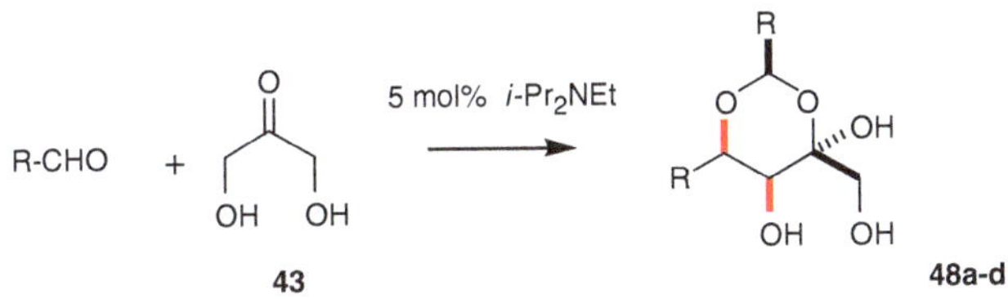

entry	R	yield (%)	*syn* / *anti*
1	*i*-Pr	92	>95 / 5
2	$PhCH_2CH_2$–	68	>95 / 5
3	Cy	94	>95 / 5
4	Ph	46	>95 / 5

Scheme 3.7.27 Amine-catalysed aldol additions to unprotected dihydroxyacetone **43**

Also, these results could be transferred successfully to aldol additions of optically active lactaldehyde and isopropylidene-protected glyceraldehyde.

An unselective reaction was observed when DBU was employed. A diastereomeric mixture of 1:1 of the corresponding rhamnulofuranose and desoxy-sorbose was detected (Scheme 3.7.28). No 1,2-asymmetric induction of the protected lactaldehyde **49** was observed. The extremely high *syn*-diastereoselectivity during

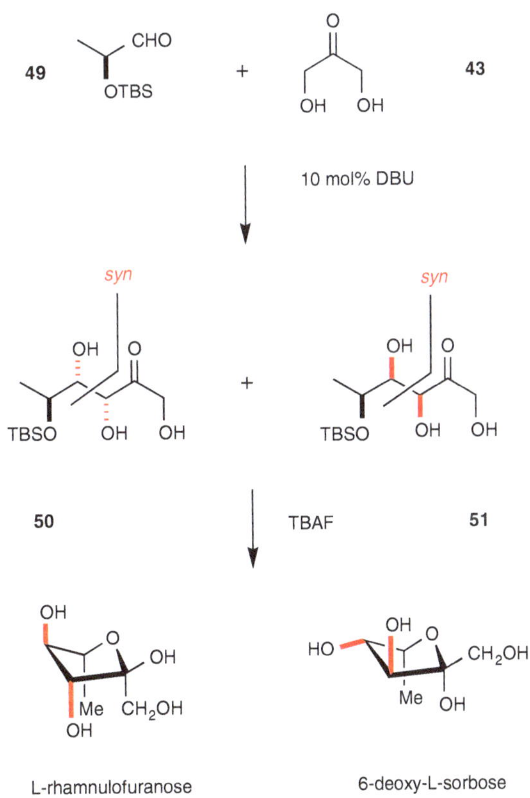

Scheme 3.7.28 *syn*-Selective amine-catalysed aldol additions of protected lactaldehyde

Scheme 3.7.29 *syn*-Selective amine-catalysed aldol additions of protected D-glyceraldehyde **52**

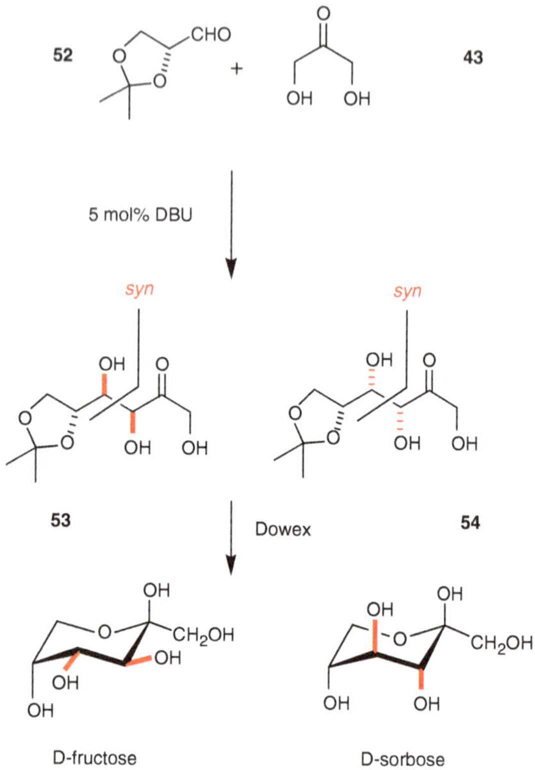

the C–C bond formation discussed above was observed again. No *anti*-configured aldol adduct could be detected.

The same results were also observed when protected D-glyceraldehyde **52** was applied in this reaction. In the presence of 5 mol% of DBU, D-fructose and D-sorbose were identified in a 1:1 mixture (Scheme 3.7.29). Similar ratios were obtained when used with other tertiary amines. By deployment of cinchonine as the tertiary amine extremely high diastereoselectivities were observed (90/10). Under these conditions the exclusive formation of fructose is observed (Scheme 3.7.30).

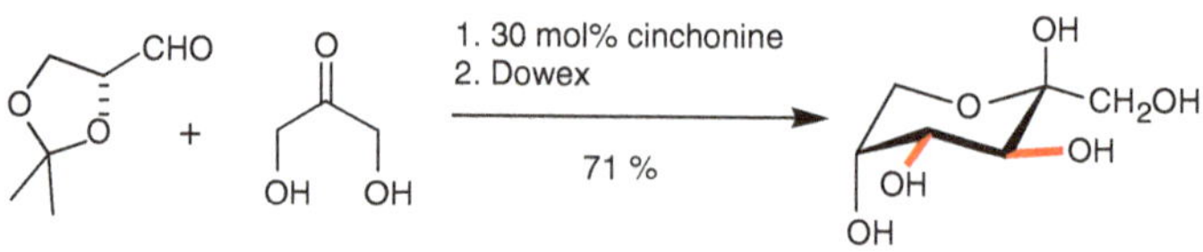

Scheme 3.7.30 Total synthesis of fructose

Based on these results the following current situation in the de novo synthesis of carbohydrates is as follows. The synthetic approach to the four ketohexoses appears to be solved by the methods described above. This can be easily accomplished by the $C_3 + C_3$ strategy for the de novo carbohydrate synthesis. With the help of

D-glyceraldehyde and protected derivatives of DHA an approach to psicose and tagatose via proline-catalysed aldol additions is given. This is due to the *anti*-preference of proline-catalysed aldol additions (Fig. 3.7.1). On the other hand fructose and sorbose are accessible – with the required *syn*-configuration – by tertiary-amine-catalysed aldol addition of DHA **43** and glyceraldehyde **52** (Scheme 3.7.31).

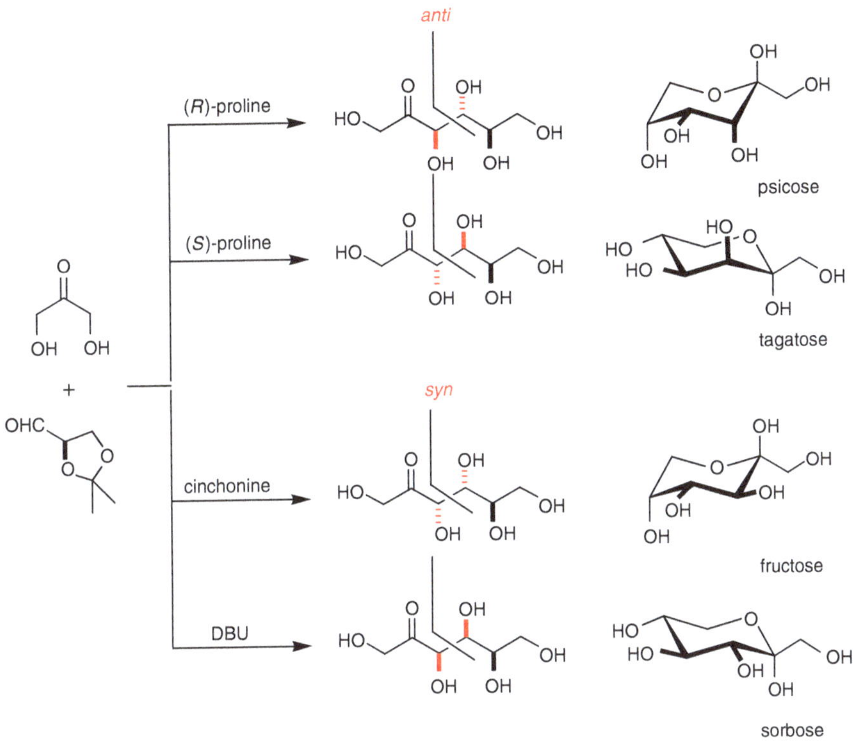

Scheme 3.7.31 Organocatalysed de novo synthesis of ketohexoses

The $C_3 + C_2$ strategy promises a synthetic access to pentoses. Very recently, Enders and Grondal described the usefulness of this concept.[96,97,98] By reacting protected DHA **41** as the C_3-unit with dimethoxyacetaldehyde **55** in the presence of substoichiometric amounts of proline protected precursors of ribose and lyxose were isolated with high degrees of enantioselectivity. Again, aldol adduct **42f** was obtained with a high degree of *anti*-diastereoselectivity under these reaction conditions. An access to *syn*-configured aldol adduct of protected DHA **41** with dimethoxyacetaldelyde **55** was reported very recently by Barbas and coworkers.[94] In these aldol transformations the authors used derivatives of threonine in substoichiometric amounts and isolated xylose precursor **46d** (Scheme 3.7.32).

Through the $C_2 + C_2 + C_2$ strategy a synthetic access to aldohexoses is given. A necessary prerequisite for a successful execution of this strategy is the defined

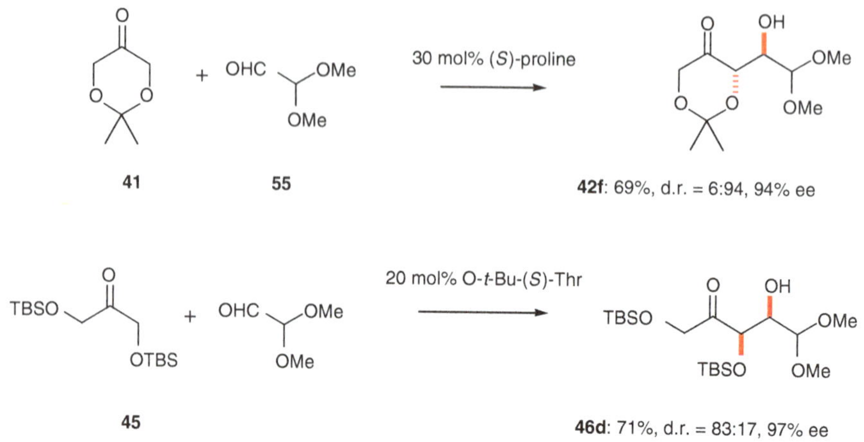

Scheme 3.7.32 $C_3 + C_2$ Approach to pentoses

and stereoselective connection of three protected glycolaldehydes. This concept was realized very recently by MacMillan and coworkers.[99] This following two-step-directed aldol addition of different aldehydes represents a temporary highlight of this development. These examples demonstrate the power of aldol methodologies available today. The authors elaborated an organocatalysed aldol addition/Mukaiyama aldol addition reaction sequence to reach this goal. By a proline-catalysed aldol addition of O-protected glycol aldehydes the *anti*-configured aldol adducts (chiral C_4-unit) were isolated with a high degree of enantioselectivity (Scheme 3.7.33).

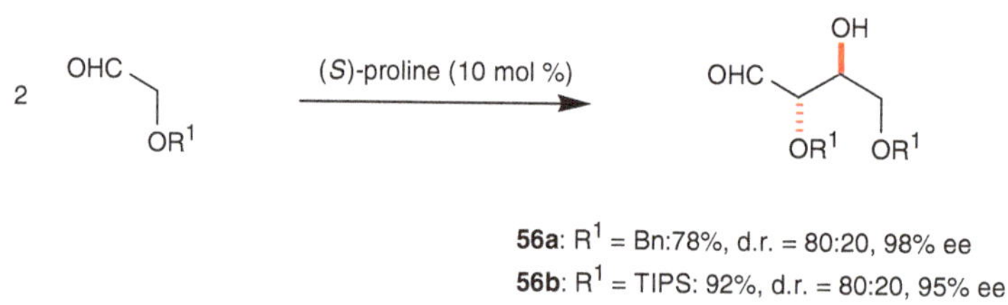

Scheme 3.7.33 (*S*)-Proline-catalysed homodimerization of oxygen-substituted acetaldehydes

Depending on conditions of subsequent stereocontrolled Mukaiyama reaction defined configured carbohydrate derivatives can be isolated with a high degree of stereoselectivity (Scheme 3.7.34). By using three different oxygen-substituted aldehydes in two separate aldol reaction steps one can obtain differently protected allose **59**, mannose **60** and glucose **61** in high yields and stereochemical purity (Scheme 3.7.34).

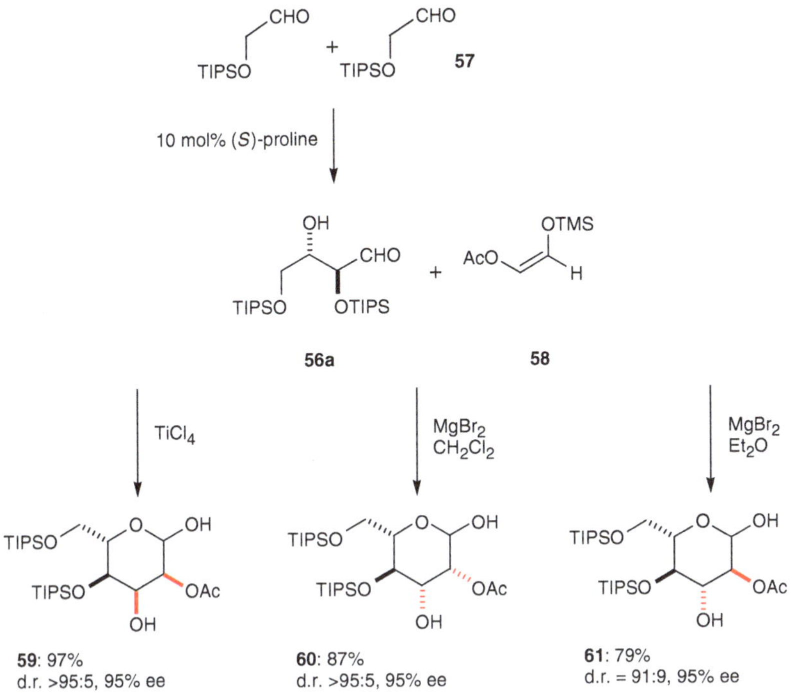

Scheme 3.7.34 Proline-catalysed/Mukaiyama aldol sequence in total synthesis of aldohexoses

By reacting enol silyl ethers of substituted aldehydes with protected hyxdroxyaldehyde **56a** under conditions of titanium(IV) -chloride catalysis different substituted and configured carbohydrates **62–65** are accessible. These results make this approach to a valuable tool in the de novo production of carbohydrates (Scheme 3.7.35).

MacMillan and coworkers described a further application of this methodology in the total synthesis of brasoside and littoralisone. During this sequence they have synthesized the required configurative defined aldol intermediate *ent*-**56a** by the strategy discussed above[100] (Scheme 3.7.36).

Pihko and coworkers applied the same aldol sequence of enolizable aldehydes in the total synthesis of prelactone B (Scheme 3.7.37).[101]

Also, carbohydrates have been synthesized by a full iterative proline-catalysed enantioselective two-step aldol addition. Córdova and coworkers obtained triketide carbohydrate **68** with an extremely high degree of enantioselectivity by reacting racemic *anti*-configurated β-hydroxyaldehydes **56a** with propionaldehyde in the presence of catalytic amounts of (*R*)-proline.[102,103,104] Thus, this method gives a highly stereoselective access to carbohydrates (Scheme 3.7.38).

For a nonselective zinc-proline-catalysed access to all eight aldohexoses using the $C_2 + C_2 + C_2$ strategy see also Kofoed et al.[105]

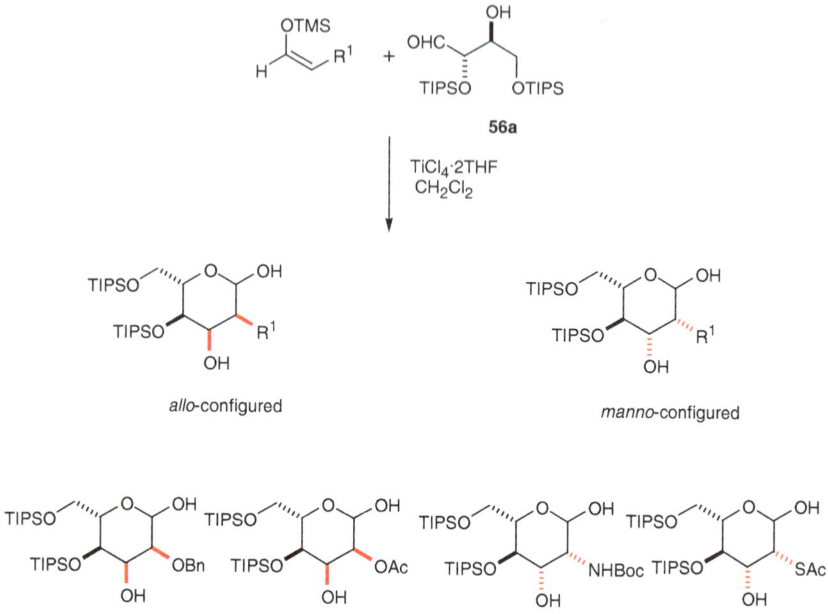

62: 83%
d.r. = 5:95, 95% ee

63: 96%
d.r.= 5:95, 95% ee

64: 74%
d.r. = 9:91, 95% ee

65: 71%
d.r. = 9:91, 95% ee

Scheme 3.7.35 Dependence of the stereochemical outcome on the nature of the protecting groups

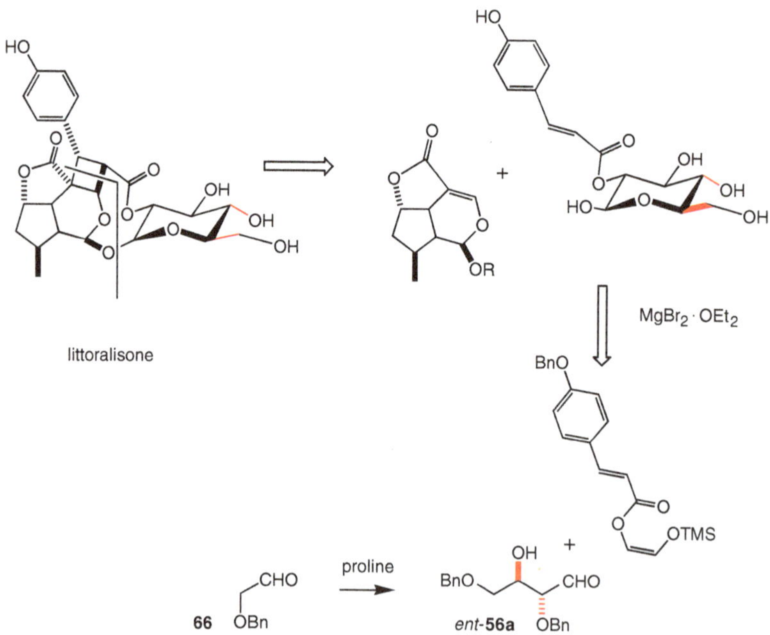

Scheme 3.7.36 Total synthesis of littoralisone

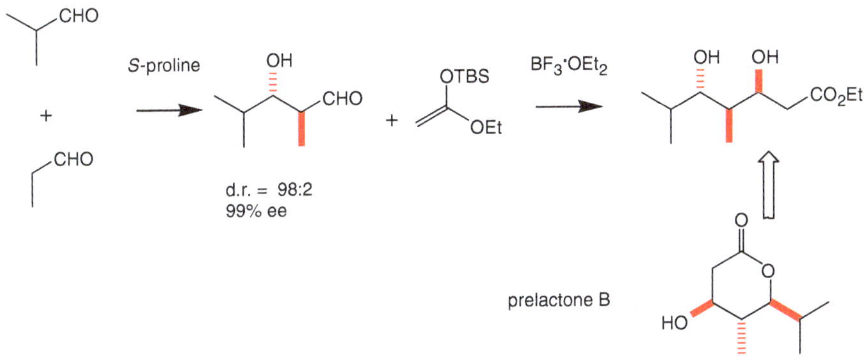

Scheme 3.7.37 Total synthesis of prelactone B

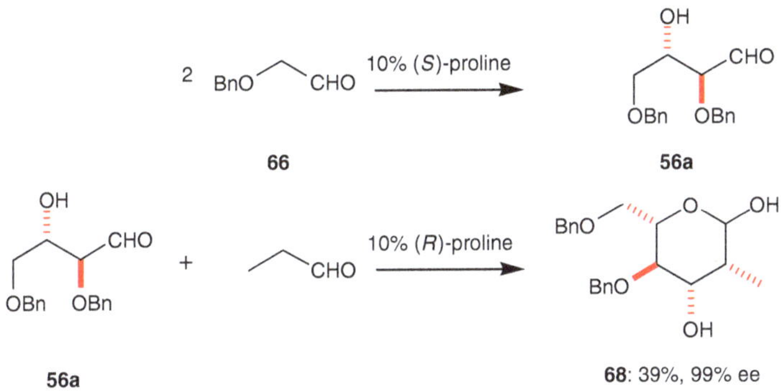

Scheme 3.7.38 Proline-catalysed two-step enantioselective approach to hexoses – de novo synthesis of carbohydrates

References

1. List B (2002) Tetrahedron 58:5572
2. List B (2004) In: Modern Aldol Additions, Mahrwald R (ed). Wiley-VCH, Weinheim, vol 1, p 161
3. Gaunt MJ, Johansson CCC, McNally M, Vo NT (2007) Drug Discovery Today 12:8
4. Bahmanyar S, Houk KN (2001) J Am Chem Soc 123:11273
5. Sevin A, Maddaluno J, Agami C (1987) J Org Chem 52:5611
6. Bahmanyar S, Houk KN, Martin HJ, List B (2003) J Am Chem Soc 125:2475
7. Fu A, Li H, Tian F, Yuan S, Si H, Duan Y (2008) Tetrahedron: Asymm 19:1288
8. Treibs W, Krumbhols K (1952) Chem Ber 85:1116
9. Schreiber J, Wermuth CG (1965) Bull Soc Chim Fr 8:2242
10. Ishikwa T, Ueda E, Okada S, Saito S (1999) Synlett 450
11. Constable EC, Harverson P, Smith DR, Whall LA (1994) Tetrahedron 50:7799
12. Mead D, Loh R, Asato AE, Liu RSH (1985) Tetrahedron Lett 26:2873
13. Lindwall HG, Maclennan JS (1932) J Am Chem Soc 54:4739

14. Buchta E, Satzinger G (1959) Chem Ber 92:449
15. Woodward RB, Sondheimer F, Taub D, Heusler K, McLamore WM (1952) J Am Chem Soc 74:4223
16. Tietze LF (1991) In: Comprehensive Organic Synthesis, Trost BM (ed). Pergamon, New York, vol 2, p 341
17. Woodward RB, Sondheimer F, Taub D, Heusler K, McLamore WM (1978) J Am Chem Soc 100:8031
18. Corey EJ, Danheiser RL, Chandasekaran S, Siret P, Keck GE, Gras JL (1978) J Am Chem Soc 100:8031
19. Snider BB, Yang K (1990) J Org Chem 55:4392
20. Sgi XX, Wu QQ, Lu X (2002) Tetrahedron: Asymm 13:461
21. Srikrishna A, Babu NC (2001) Tetrahedron Lett 42:4913
22. Traber B, Pfander H (2000) Tetrahedron Lett 41:7197
23. Roush WR, Barda DA (1997) Tetrahedron Lett 38:8785
24. Lalande R, Moulines J, Duboudin JCR (1962) Bull Soc Chim Fr 10987
25. Wei CQ, Jiang XR, Ding Y (1989) Tetrahedron 54:12623
26. Allard M, Levisalles J, Rudler H (1968) Bull Soc Chim Fr 303
27. Wright J, Drtina GJ, Roberts RA, Paquette LA (1988) J Am Chem Soc 110:5806
28. Wettstein A, Heusler K, Ueberwasser H, Wieland P (1957) Helv Chim Acta 40:323
29. Begbie AL, Golding BT (1972) J Chem Soc Perkin I 602
30. Corey EJ, Tius M, Das J (1980) J Am Chem Soc 102:7612
31. Swaminathan S, Newman MS (1958) Tetrahedron 2:88
32. Selvarajan R, John JP, Narayanan KV, Swaminathan S (1966) Tetrahedron 22:949
33. Eder U, Sauer G, Wiechert R (1971) Angew Chem Int Ed Engl 10:496
34. Hajos ZG, Parrish DR (1974) J Org Chem 39:1615
35. Hayashi Y, Sekizawa H, Yamaguchi J, Gotoh H (2007) J Org Chem 72:6493
36. Barbas CF III (2008) Angew Chem Int Ed 47:42
37. Danishefsky S, Cain P (1976) J Am Chem Soc 98:4975
38. Agami C, Meynier F, Puchot S (1984) Tetrahedron 40:1031
39. Hayashi Y, Itoh T, Nagae N, Ohkubo M, Ishikawa H (2008) Synlett 1565
40. Hajos ZG, Parrish DR (1974) J Org Chem 39:1615
41. Agami C, Puchot S (1986) Tetrahedron 42:2037
42. Agami C, Levisalles J, Puchot S (1985) Chem Commun 441
43. Rajapogal D, Moni MS, Subramanian S, Swaminathan S (1999) Tetrahedron: Asymm 10:1631
44. Bahmanyar S, Houk KN (2001) J Am Chem Soc 123:12911
45. Allemann C, Gordillo R, Clemente FR, Cheong PHY, Houk KH (2004) Acc Chem Res 37:558
46. Gröger H, Wilken J (2001) Angew Chem Int Ed 40:529
47. List B (2004) Acc Chem Res 37:548
48. Guillena G, Najera C, Ramon DJ (2007) Tetrahedron: Asymm 18:2249
49. Berkessel A, Gröger H (eds) (2005) Asymmetric Organocatalysis, Wiley-VCH, Weinheim
50. Mukherjee S, Yang JW, Hoffmann S, List B (2007) Chem Rev 117:5471
51. Tanaka F, Barbas CF III (2007) In: Enantioselective Organocatalysis, Dalko P (ed). Wiley-VCH, Weinheim, p 19
52. List B (2006) Chem Commun 819
53. Dondoni A, Massi A (2008) Angew Chem Int Ed Engl 47:4638
53. Pidathala C, Hoang L, Vignola N, List B (2003) Angew Chem Int Ed Engl 42:2785
55. Mans DM, Pearson WH (2004) Org Lett 6:3305
56. Chandler CL, List B (2008) J Am Chem Soc 130:6737
57. List B, Lerner RA, Barbas CF III (2000) J Am Chem Soc 122:2395
58. Notz W, List B (2000) J Am Chem Soc 122:7386
59. Saito S, Nakadai M, Yamamoto H (2001) Synlett 8:1245
60. Cordova A, Weibaiao Z, Ibrahem I, Reyes E, Engqvist M, Liao WW (2005) Chem Commun 3586

61. Jiang Z, Liang Z, Wu X, Lu Y (2006) Chem Commun 2801
62. Mase N, Tanaka F, Barbas III CF (2004) Angew Chem Int Ed 43:2420
63. Mase N, Tanaka F, Barbas III CF (2003) Org Lett 5:4369
64. Northtrup AB, MacMillan DWC (2002) J Am Chem Soc 124:6798
65. Kumaraswamy G, Markondaiah B (2007) Tetrahedron Lett 48:1707
66. (a) Wu Y, Zhang Y, Yu M, Zhao G, Wang S (2006) Org Lett 8:4417; (b) Tanimori S, Naka T, Kirihata M (2004) Synth Commun 34:4043; (c) Tanimori S, Naka T, Kirihata M (2004) Synth Commun 34:4043; (d) Guillena G, Hita MC, Najera C (2006) Tetrahedron: Asymm 17:1027; (e) Ma GN, Zhang YP, Shi M (2007) Synthesis 197; (f) Wang W, Li H, Wang J (2005) Tetrahedron Lett. 46:5077; (g) Guo HM, Cun LF, Gong LZ, Mi AQ, Jiang YZ (2005) Chem Commun 1450; (h) Tang Z, Yang ZH, Chen XH, Cun LF, Mi AQ, Jiang YZ, Gong LZ (2005) J Am Chem Soc 127:9285; (i) He L, Tang Z, Cun LF, Mi AQ, Jiang YZ, Gong LZ (2006) Tetrahedron 62:346; (k) Jiang M, Zhu SF, Yang Y, Gong LZ, Zhou XG, Zhou QL (2006) Tetrahedron: Asymm 17:384; (l) Lacoste E, Landais Y, Schenk K, Verlhac JB, Vincent JM (2004) Tetrahedron Lett 45:8035; (m) Berkessel A, Koch B, Lex J (2004) Adv Synth Catal 346:1141; (n) Silva F, Sawicki M, Gouverneur V (2006) Org Lett 8:5417; (o) Chen JR, Lu HH, Li XY, Cheng L, Wan J, Xiao WJ (2005) Org Lett 7:4543; (p) Tang Z, Yang ZH, Cun LF, Gong LZ, Mi AQ, Jiang YZ (2004) Org Lett 6:2285; (q) Gryko D, Lipinski R (2005) Adv Synth Catal 347:1948; (r) Cheng C, Sun J, Wang C, Zhang Y, Wei S, Jiang F, Wu Y (2006) Chem Commun 215; (s) Gu L, Yu M, Wu X, Zhang Y, Zhao G (2006) Adv Synth Catal 348:2223; (t) Kano T, Takai J, Tokuda O, Maruoka K (2005) Angew Chem Int Ed 44:3055; (u) Hayashi Y, Itoh T, Aratake S, Ishikawa H (2008) Angew. Chem. Int. Ed. 47:2082; (v) Mei K, Zhang S, He S, Li P, Jin M, Xue F, Luo G, Zhang H, Song L, Duan W, Wang W (2008) Tetrahedron Lett 49:2681
67. Tang Z, Marx A (2007) Angew Chem Int Ed 46:7297
68. Sun G, Fan J, Wang Z, Lin Y (2008) Synlett 2491
69. Oberhuber M, Joyce GF (2005) Angew Chem Int Ed 44:7580
70. Benaglia M, Celentano G, Cozzi F (2001) Adv Synth Catal 343:171
71. (a) Darbre T, Machuqueiro M (2003) Chem Commun 9:1090; (b) Kofoed J, Machuqueiro M, Reymond JL, Darbre T (2004) Chem Commun 13:1540; (c) Torii H, Nakadai M, Ishihara K, Saito S, Yamamoto H (2004) Angew Chem Int Ed 43:1983; (d) Nyberg AI, Usano A, Pihko PM (2004) Synlett 1891; (e) Hayashi Y, Sumiya T, Takahashi J, Gotoh H, Urushima T, Shoji, M (2006) Angew Chem Int Ed 45:958; (f) Hayashi Y, Aratake S, Okano T, Takahashi J, Sumiya T, Shoji M (2006) Angew Chem Int Ed 45:5527; (g) Amedjkouh M (2005) Tetrahedron: Asymm 16:1411; (h) Dziedzic P, Zou W, Hafren J, Córdova A (2006) Org Biomol Chem 4:38; (i) Kofoed J, Darbre T, Reymond JL (2006) Chem Commun 14:1482; (k) Chimni SS, Mahajan D (2006) Tetrahedron: Asymm 17:2108; (l) Peng YY, Ding QP, Li Z, Wang PG, Cheng JP (2003) Tetrahedron Lett 44:3871; (m) Cobb AJA, Shaw DM, Ley SV (2004) Synlett 558; (n) Mlynarski J, Paradowska J (2008) Chem Soc. Rev 37:1502; (o) Amedjkouh M (2007) Tetrahedron: Asymm 18:390; (p) Liu K, Häussinger D, Woggon WD (2007) Synlett 2298
72. Bogevig A, Kumaragurubaran N, Jørgensen KA (2002) Chem Commun 620
73. Bogevig A, Poulsen TB, Zhuang W, Jørgensen KA (2003) Synlett 1915
74. Luo S, Xu H, Chen L, Cheng JP (2008) Org Lett 10:1775
75. Mase N, Nakai Y, Ohara N, Yoda H, Takabe K, Tanaka F, Barbas CFIII (2006) J Am Chem Soc 128:734
76. Zheng Y, Avery MA (2004) Tetrahedron 60:2091
77. Nicolaou KC, Ninkovic S, Sarabia F, Vourloumis D, He D, Vallberg H, Finlay MRV, Zang Z (1997) J Am Chem Soc 119:7960
78. List B, Pojarlioev P, Castello C (2001) Org Lett 3:573
79. Zou B, Wie J, Cai G, Ma D (2003) Org Lett 5:3503
80. Ma D, Zou B, Cai G, Hu Y, Liu JO (2006) Chem Eur J 12:7615
81. Zhang S, Duan W, Wang W (2006) Adv Synth Cat 348:1228

82. Luppi G, Monari M, Correa RJ, Violante FdeA, Pinto AC, Kaptein B, Broxterman QB, Garden SJ, Tomasini C (2006) Tetrahedron 62:12017
83. Itagaki N, Iwabuchi Y (2007) Chem Commun 1175
84. (a) Itagaki N, Kimura M, Sugahara T, Iwabuchi Y (2005) Org Lett 7:4185; (b) Itagaki N, Sugahara T, Iwabuchi T (2005) Org Lett 7:4181
85. Sun B, Peng L, Chen X, Li Y, Li Y, Yamasaki K (2005) Tetrahedron: Asymm 16:1305
86. Enders D, Palecek J, Grondal C (2006) Chem Commun 655; (b) Grondal C, Enders D (2006) Synlett 3507; (c) Enders D, Gasperi T (2007) Chem Commun 88
87. Enders D, Terteryan V, Palecek J (2008) Synthesis 2278
88. Kumar I, Rode CV (2007) Tetrahedron: Asymm 18:1975
89. Córdova A, Notz W, Barbas CFIII (2002) Chem Commun 3024
90. for selected references, see: (a) Sakthiviel K, Notz W, BuiT, Barbas CFIII (2001) J Am Chem Soc 123:5260; (b) Magiotti V, Resmini M, Gouverneur V (2002) Angew Chem 114:1064; (c) Liu H, Peng L, Zhang T, Li Y (2003) New J Chem 27:1159; (d) Magiotti V, Bahmanyar S, Reiter M, Resmini M, Houk KN, Gouverneur V (2004) Tetrahedron 60:619; (e) Pan Q, Zou B, Wang Y, Ma D (2004) Org Lett 6:1009; (f) Tang Z, Yang ZH, Cun LF, Gong LZ, Mi AQ, Jiang YZ (2004) Org Lett 6:2285; (g) Storer RI, MacMillan DWC (2004) Tetrahedron 60:7705; (h) Samanta S, Liu J, Doddam R, Zhao CG (2005) Org Lett 7:5321; (i) Hayashi Y, Sumiya T, Takahashi J, Gotoh H, Urushima T, Shoji M (2006) Angew Chem 118:972; (k) Guillena G, Hita MdC, Najera C (2006) Tetrahedron: Asymm 17:1027; (l) Gu Q, Wang XF, Wang L, Wu XY, Zhou XL (2006) Tetrahedron: Asymm 17:1537; (m) Calderon F, Doyaguez EG, Ferrnandez-Mayoralas A (2006) J Org Chem 71:6258; (n) Ramasastry SSV, Zhang H, Tanaka F, Barbas CFIII (2007) J Am Chem Soc 129:288; (o) Chen XH, Luo SW, Tang Z, Cun LF, Mi AQ, Jiang YZ, Gong LZ (2007) Chem Eur J 13:689; (p) Guillena G, Hita MdCM, Najera C (2007) Tetrahedron: Asymm 18:1272; Luo S, Xu H, Zhang L, Li J, Cheng JP (2008) Org Lett 10:653
91. for selected references see: (a) Cordova A, Zou W, Ibrahem I, Reyes E, Engquvist M, Liao WW (2005) Chem Commun 3586; (c) Zou W, Ibrahem I, Dziedzic P, Sunden H, Cordova A (2005) Chem Commun 4946; (d) Westermann B, Neuhaus C (2005) Angew Chem Int Ed 44:4077; (e). Suri T, Ramachary DB, Barbas CFIII (2005) Org Lett 7:1383; (f) Fernandez-Lopez R, Kofoed J, Machuqueiro M, Darbre T (2005) Eur J Org Chem 24:268; (g) Suri FT, Mitsumori S, Albertshofer K, Tanaka F, Barbas CFIII (2006) J Org Chem 71:3822; (h) Ibrahem I, Zou W, Xu Y, Cordova A (2006) Adv Synth Catal 348:211; (i) Cordova A, Zou W, Dziedzic P, Ibrahem I, Reyes E, Xu Y (2006) Chem Eur J 12:5383; (k) Grondal C, Enders D (2006) Tetrahedron 62:329; (l) Majewski M, Niewczas I, Palyam N (2006) Synlett 2387; (m) Hayashi Y, Aratake S, Itoh T, Okano T, Sumiya T, Shoji M (2007) Chem Commun 957; (n) Luo S, Mi X, Zhang L, Liu S, Xu H, Cheng JP (2007) Tetrahedron 63:1923; (o) Grondal C, Enders D (2006) Synlett 3597
92. Enders D, Grondal C (2005) Angew Chem Int Ed 44:1210
93. Ramasastry SSV, Albertshofer K, Utsumi N, Tanaka F, Barbas CFIII (2007) Angew Chem Int Ed 46:5572
94. Utsumi N, Imai M, Tanaka F, Ramasastry SSV, Barbas CFIII (2007) Org Lett 9:3445
95. Markert M, Mulzer M, Schetter B, Mahrwald R (2007) J Am Chem Soc 129:7258
96. Enders D, Grondal C (2005) Lett Org Chem 2:577
97. Enders D, Palecek J, Grondal C (2006) Chem Commun 655
98. Grondal C, Enders D (2007) Adv Synth Catal 349:694
99. (a) Northrup AB, Mangion IK, Hettche F, MacMillan DWC (2004) Angew Chem Int Ed 43:2152; (b) Northrup AB, MacMillan DWC (2004) Science 305:1752; (c) Mangion IK, Northrup AB, MacMillan DWC (2004) Angew Chem Int Ed 43:6722
100. Mangion IK, MacMillan DWC (2005) J Am Chem Soc 127:3696
101. Pihko PM, Erkkila A (2003) Tetrahedron Lett 44:7607
102. Reyes E, Córdova A (2005) Tetrahedron Lett 46:6605

103. Casas J, Engqvist M, Ibrahem I, Kaynak B, Córdova A (2005) Angew Chem Int Ed Engl 44:1343
104. Córdova A, Ibrahem I, Casas J, Sundén H, Engqvist M, Reyes E (2005) Chem Eur J 11:4772
105. Kofoed J, Reymond JL, Darbre T (2005) Org Biomol Chem 3:1850

Index

FSC
www.fsc.org
MIX
Papier | Fördert
gute Waldnutzung
FSC® C083411

Kolibri 360 GmbH
Ferdinand-Jühlke-Straße 7
99095 Erfurt, Deutschland
produktsicherheit@kolibri360.de